International construction

General Editor: Colin Bassett BSc, FCIOB, FFB

International construction

Marketing, planning and execution

Victor L. Cox

Construction Press
London and New York

Construction Press
an imprint of:
Longman Group Limited
Longman House, Burnt Mill, Harlow
Essex CM20 2JE, England
Associated companies throughout the world.

*Published in the United States of America
by Longman Inc., New York*

First published 1982

British Library Cataloguing in Publication Data
Cox Victor L.
   International construction: marketing, planning
  and execution.
   1. Construction industry — Great Britain
   2. Export marketing
   I. Title
   624′. 068′8   HD9715.G72
   ISBN 0-582-30510-1

Library of Congress Cataloging in Publication Data
Cox, Victor L., 1928–
   International construction.

   Includes index.
   1. Construction industry. I. Title.
   HD9715.A2C69   624′.068   82-1433
   ISBN 0-582-30510-1   AACR2

Printed in Great Britain by
The Pitman Press Ltd., Bath

Contents

Introduction

Construction output in the UK rose from about £18.7 bn in 1979 to in excess of £21.2 bn in 1981, but output at 1975 prices fell from £11.5 bn to £9.6 bn. This and trading objectives generally have led many consultants, manufacturers and contractors, like those elsewhere in the European Economic Community and most industrial countries who have faced similar conditions, to look for growth in other international construction markets.

If developing nations and others are to progress in any way towards the standards achieved by leading industrial countries construction will be essential. They may expect their populations to grow by more than 500 m. in the next twenty years creating demands for construction to support this increase, but generally those with the largest demand will be least able to afford the work needed and are less likely in the short term to have all the trade or management skills necessary. Such developments could establish new trends. Although there will still be substantial projects, smaller projects offering greater management flexibility and in keeping with financial limits are more probable. Finding employment for growing populations will also encourage this. Markets for the commercial, technical and management skills available from the UK construction industry could increase, and consultants, products manufacturers and contracting firms may find it necessary to pursue these by providing manufacturing expertise, technology and training.

Organizations of varying sizes from the UK have made great contributions to world construction over many years and to the UK balance of payments. The industry and particular firms are, as a result highly regarded. In the financial year to 31 March 1981 overseas earnings by the industry were £3.18 bn made up of £500 m. by consultants, £2.13 bn by manufacturers and suppliers and £550 m. by contractors. The geographical spread of contracts obtained and earnings in 1981 and in the preceding two financial years show the extent of success and involvement by the industry. Details are given in Table 0.1.

Competition is, nevertheless, severe and unlikely to lessen. It has a number of facets. The most significant is that UK contractors, like many others, particularly in Western Europe, must often compete on unequal terms. As an example, some organizations receive considerable government support, and their entry into a particular market may be governed by the need to obtain hard currency or

Table 0.1 Overseas construction contracts obtained and earnings 1978–81

	1978–79	1979–80	1980–81
Geographical spread and value of contracts (£m.)			
European Economic Community	39	27	39
Rest of Europe	75	82	81
Middle East (Asia)	632	553	358
Middle East (Africa)	45	100	54
Rest of Asia	195	89	132
Rest of Africa	229	203	348
America	91	274	236
Oceania	81	58	112
All countries	1 387	1 385	1 360
Construction industry earnings (£m.)			
Designers	460	470	500
Manufacturers and suppliers	1 900	2 030	2 130
Contractors	345	240	550
Total	2 705	2 740	3 180

for other reasons. Others may enter markets as a part of a development plan and rely on government planning, manpower support and financial inducements. Their costs can be such that it is difficult for UK firms to compete against them in some markets. However, all are finding that worldwide competition creates pressures, and none is immune to world inflationary trends.

Multilateral and bilateral aid and regional trading blocks could exert increasing influences on competition. In many growing markets much will depend on the availability of finance. Some countries which seem to have potential and opportunities to provide or obtain finance may also have experience or developing construction industries of their own, while those with very considerable markets often must work within the constraints of limited availability of finance, political uncertainties and little industry. In the UK industry could become increasingly influenced by the EEC, which may also be expected to influence some overseas construction markets, and the Commission of the EEC plans to harmonize some construction activities of member states and also wishes to harmonize trading. Company law, customs procedures and export credits are examples.

Apart from the massive bilateral aid given by the UK and other members of the EEC on a national basis, the EEC is a growing force in world construction through funds made available for projects as a part of the European Development Fund. The UK construction industry can share in contracts financed by this aid.

This book is about trading conditions, trends and methods of assessing and entering international construction markets. It has been written in relation to UK procedures and facilities, but is also based on principles which could be applicable in other countries. The framework will, it is hoped, provide those new to international construction, whether designers, manufacturers, contractors or from other disciplines, with essential information on which to base analysis and decisions. For those already familiar with what is involved the information given may provide reminders and methods of working. Those involved in construction, but who are essentially a part of another discipline, such as accountants, bankers, insurers, lawyers and tax specialists, may find sections of the book relevant to their activities and a means of integrating their skills into a broader industry framework.

Some statistics have been used. While these quickly become out of date they contribute to a method of working. The principles of their use should not necessarily change and, in any event, up-dating would usually be a necessary part of any market assessment. Sections of the book refer to some legal, taxation and similar matters on which there will always be a need for skilled guidance from those dealing with such subjects at the time of assessment and decision-taking. They are liable to frequent change which can have a far-reaching effect on a course of action, but with these, too, every attempt has been made to provide a framework within which to work.

References are made, depending on the subject, to the customer, buyer, client or employer. While legal definitions are likely to be important in each case and in any contract for the purpose of assessing markets they are similar. The essential definition is based on the ability and authority to initiate and pay for the construction design, product or contract being considered, and it is on this that a market decision will succeed or fail in any country or sector of the construction industry. Throughout, references to consultants or designers include any relevant professionally qualified group or individuals who may be expected to give advice and prepare feasibility studies as well as prepare designs. Other descriptions and roles such as those for manufacturers and contractors will generally be evident from the context used.

My thanks and appreciation are due to my wife, without whose help this book would probably have never been prepared, and to many friends and colleagues in the UK and elsewhere whose information, guidance and cooperation over a number of years have ultimately contributed to what has been written.

Victor L. Cox
1 July 1982

Chapter 1

UK construction industry. Structure, control and research

Value of comparisons

For those planning overseas construction work a detailed understanding of the structure of the home construction industry or that of a major industrial country is invaluable. Many will have such knowledge, but they may also need to undertake a conscious reappraisal. Much overseas construction is frequently undertaken in countries with a developing or partially developed industrial structure. Those already active overseas, or responsible for organizing or contributing to construction work, should find comparisons helpful.

The UK construction industry may have many critics, but it is effective and has developed from a basis of free-enterprise associated with government planning over a very long period. It has a structure which overseas countries may find useful to emulate or adapt to their own needs.

Output

Construction output in the UK is paid for by public and private sector finance. Emphasis alters with changes of governments. In recent years the balance was similar under any administration, however, the current emphasis is on private sector expenditure. The UK is a highly industralized country and has a population of about 55 m. Urban development is considerable and population densities are high. Expenditure on construction in 1981 amounted to over £21.2 bn of which about £5.0 bn was for public sector new work, £7.6 bn for new private contracts and repair and maintenance over £8.5 bn but the apparent growth compared with previous years is due to inflation. Expenditure at constant prices shows a fall reflecting current government policies and market conditions and also a relatively high satisfaction of needs which are in common with other industrialized countries. Statistics for 1979 to 1981 are shown in Table 1.1.

Manpower

More than 20 000 architects practise in the UK, with about 9 000 quantity surveyors and 18 000 electrical, mechanical and structural engineers. It is estimated

1

Table 1.1 Construction industry output in the UK (£m.)

	1979	1980	1981
Housing			
Public	1 747	1 753	1 234
Private	2 731	2 652	2 553
Other work			
Public	3 285	3 785	3 789
Private			
Industrial	2 402	2 879	2 417
Commercial	1 961	2 495	2 699
Repairs and maintenance	6 659	8 473	8 559
Total (current prices)	18 785	22 037	21 251
Total (1975 prices)	11 511	10 923	9 674
Index (1975 = 100)	100.8	95.6	84.6

that approximately 12 000 materials and components manufacturers supply products to the construction industry at home and abroad. Actual construction is undertaken by over 70 000 contracting and sub-contracting organizations. Of these, only about 50 employ more than 1 200 men. Total manpower employed by the industry in 1981 was about 1.1 m. compared with about 3.7 m. employed in engineering and total UK manpower employed of over 26.0 m.

Economic planning

It can be seen that the UK construction industry can exert considerable influence on economic planning, on industrial output and above all on social developments. Although projects are initiated by the public and private sectors, output by the public sector is subject to much more unified control than private sector work. Planning by goverment and other public sector departments is, therefore, of crucial influence on the industry. Any change of government policy can have severe repercussions. All too often it is used as an economic regulator to the detriment of output and balanced progress.

The Department of the Environment (DoE) is the government department which exerts most immediate influence on the construction industry, but is subject to the financial control of the Treasury. The DoE is headed by a Secretary of State, under whom are ministers for local government and development, transport industries and for the housing and construction industries. Work of the Department is administered through a central and regional organization. A number of directorates are responsible for particular construction matters. A Department of Health and Social Security is responsible for health building programmes and the Department of Education and Science is responsible for school and university building. Central government policies can be administered directly through government departments, but in many cases they are im-

2

plemented through local authorities. The Greater London Council and the metropolitan counties and districts have by far the greatest autonomy, followed by county districts and parish councils. For some purposes, such as the allocation of development funds, regional areas have been established, but local authorities are, nevertheless, paramount within these regions.

Structure

Specification and purchase is undertaken by numerous organizations. In each, professional designers have considerable influence. A consortia approach has been adopted in many cases in the past to encourage effective buying policies. Regional hospital boards are responsible for health buildings. More than fifty housing consortia have existed to carry out public sector housing developments and schools consortia were created to provide for educational buildings. The latter was probably the most successful. In addition, large public corporations responsible for railways, gas and electricity supplies and a variety of other undertakings exert their influence. Private sector buying patterns are more fragmented, but are often more effective.

Design professions and those responsible for legal or financial control are usually the responsibility of chartered institutes. Those affecting the construction industry are the Royal Institute of British Architects (RIBA), the Royal Institution of Chartered Surveyors (RICS) and the Institution of Civil Engineers (ICE). Others concerned with all types of design and engineering are equally important and support most aspects of construction. Organizations such as the National Federation of Building Trades Employers (NFBTE), the Federation of Civil Engineering Contractors (FCEC) and similar organizations have an important position within the industry as bodies that establish pay agreements with trade unions. They also advise government departments on industry views and act as a means of consultation between the two sectors. Trade associations represent manufacturers of materials and many products and encourage the maintenance of standards. They range over a very comprehensive field of activity. The National Council of Building Material Producers (BMP) is the leading body concerned with construction products. Research organizations are equally comprehensive. The Building Research Establishment (BRE) under the control of the DoE, is important. Several trade unions operate throughout the construction industry. Manufacturers can be influenced by a very considerable number, but contractors and those concerned with sub-contracting are generally most influenced by the Transport and General Workers' Union (TGWU) and the Union of Construction, Allied Trades and Technicians (UCATT). Many of the organizations have an information service. Key sources of information in the UK are several building centres, the most comprehensive of which is The London Building Centre. There are also several highly developed information guides, particularly for products.

Functioning of the National Building Agency (NBA) was an example of the

UK public and private sector mixed economy which is so much a feature of UK industry, and the construction industry in particular. The NBA was formed by the Minister of Public Building and Works, now part of the DoE, in 1964, and earlier objectives were to act as an independent advisory body to promote the use of improved design, management and site operation techniques in the public and private sectors. For several years there was emphasis on industrialization and use of building systems, and much was done for and through local authorities. The NBA introduced an appraisal certificate for systems which encouraged their acceptance throughout the construction market. Later the role of the organization covered wider public and private sector activities, particularly consultancy to supplement finance provided from government sources. It has now ceased activities as there are ample private sector facilities available.

Studies and reports

Research is an important part of construction in the UK. Development of new materials and products is essential and is part of the complex nature of UK industrial progress. It covers design, management and assembly techniques. Legislation and regulations may follow as a part of government and industry aims to improve output, methods, quality and safety. In the past twenty-five years the role of the construction industry within the national economy and the relationship between the public and private sectors has led to a number of studies and reports. These were designed to improve the structure of the industry and to comment on many changes within the industry. One of the most important was the Parker Morris Report in 1961 which set out minimum desirable criteria of accommodation, heating, insulation, fittings and environment for mass housing. Others surveyed problems facing the construction industry, placing and management of contracts for building and civil engineering work and regional needs. The Economic Development Committee for Building and the Economic Development Committee for Civil Engineering, which are responsible to the National Economic Development Office (NEDO) comment on industry markets. The National Economic Development Office prepares output forecasts from time to time and is also a means of consultation in the industry and with government planners.

Reports have been prepared on matters concerning manpower in the building and civil engineering industries, operatives in the building industry, housing improvement, architects' costs and fees, selective tendering for local authorities, technicians' courses and examinations, small firms, registration of builders, building maintenance and a variety of other subjects. They show clearly the high degree of study which has been carried out in the construction industry in the UK and the complexity of matters which need to be understood by those working within it or wishing to enter from elsewhere.

Legislation

Legislation which influences construction in the UK is also extensive and very complex. Detailed professional advice is usually needed for each design, manufacturing or assembly operation. The Building Control Act, 1966 was introduced to regulate building and constructional work. This is still on the Statute Book, but building is generally unrestricted in relation to size and frequency although there can be many other restrictions. The Clean Air Act, 1956 provides for abating the pollution of the air. Further provisions were made in an Act passed in 1968. During 1967–72 government cost yardsticks for public sector housing were introduced. Much has been done to restrict the uncontrolled growth of urban developments within rural areas. The Countryside Act, 1968 was particularly important. Manufacturers are affected by the Factories Act, 1961 which relates to the safety, health and welfare of employed persons. Manufacture of products is influenced by the provisions of the Fire Precautions Act, 1971. European Economic Community (EEC) legislation is now also of growing importance.

In 1969 the Housing Act made provisions for grants by local authorities and government contributions towards the cost of providing dwellings by conversion or improvement of dwellings or houses. This was a move to meet rising costs for new housing and as a means of retaining older buildings. Modifications have been made. Emphasis on the need to improve the environment is increasing and as far back as 1960 an Act was passed to make new provisions in respect of the control of noise and vibration. In 1957 an Act relating to the thermal insulation of buildings was passed. Town planning legislation, and Town and Country Planning Acts influence development and the control of the environment in the greatest way.

Another major influence is the use of building regulations. Design of buildings and specification of building products is governed by the Building Regulations, 1972 which can refer to British Standards and Codes of Practice. In turn, local authorities can pass their own bye-laws. The Greater London Council and the Manchester City Council legislation influence large areas of construction activity.

The Building Regulations, 1972 are in several parts: part A – general; part B – materials; part C – preparation of site and resistance to moisture; part D – structural stability; part E – structural fire precautions; part F – thermal insulation; part G – sound insulation; part H – stairways and balustrades; part J – refuse disposal; part K – open spaces, ventilation and height of rooms; part L – chimneys, flue pipes, hearths and fireplace recesses; part M – heat-producing appliances and incinerators; part N – drainage, private sewers and cesspools; part P – sanitary conveniences, and part Q – ashpits, wells, tanks and cisterns. The Public Health Act, 1936, the Clean Air Act, 1956 and the Fire Precautions Act, 1971 give authorities for these regulations. Regulations are also influenced by other legislation mentioned earlier and interpretation can be exceedingly difficult. Their application is controlled by building inspectors, but the DoE Building Regulation Advisory Committee (BRAC) representing government departments and industry has a key role. It could be replaced by a Building Control Council.

Specification

The construction industry structure which has developed in the UK, while heavily influenced by government legislation, is effective largely as a result of ways in which all sectors of the industry respond to design and specification of construction products, codes of practice and other influences on output.

British Standards are prepared by the British Standards Institution (BSI). This is a national body. It is financed by government and industry subscriptions and through a very comprehensive committee structure, codes of practice and standards for materials and products and design disciplines can be agreed. Products which are accepted through long usage or because they have a British Standard can be readily acceptable to specifiers. Products which for some reason cannot meet these conditions have been tested against given criteria by the Agrément Board, and in this way found acceptance. The Fire Officers Committee (FOC) which was established by independent insurers exerts very considerable authority when establishing insurance premiums for buildings. Decisions taken can influence design and use of materials and products. Work of the FOC must be seen in relation to the Building Regulations, 1972 and fire regulations generally. Fire tests on building materials and structures have priority in the UK. They are given in a British Standard and sales depend upon them.

Although there are some changes in the role of the architect as a specifier, influence exerted by the profession is still paramount. Architects employed by public sector departments and in contracting and developers' organizations are particularly influential. In the UK the role of the quantity surveyor is important in relation to design, costing and the eventual control of the use of specified products and payment by measurement of work done. The widest possible range of materials and products is manufactured in the UK. Manufacture may be to British Standards or as a result of common acceptance or to meet the requirements of an Agrément certificate.

Selling is generally undertaken through the sales forces of manufacturers or through agents or licensees or other distributors in a similar way to other industrial countries. Products for major contracts may be supplied direct to a large user, but nevertheless the role of the builders' merchant in the UK is still an active one. Until recently there were numerous builders' merchants supplying heavy construction materials, large and small components, protective materials and either stocking or taking orders for the widest range of goods. There are still many merchants, but mergers have led to operations on a larger scale, greater efficiency, and, in some cases, more specialization.

Main contractors and the way in which they undertake general contracting is highly developed in the UK and there is a very wide network of sub-contractors for all sectors of the industry. Work is governed by a Standard Form of Building Contract. The RIBA has introduced a Form of Warranty for Nominated Sub-Contractors. Education and training is well organized through schools and universities and through professional and trade bodies. Through legislation a Construction Industry Training Board was established several years ago. The industry always needs craftsmen, and organizations such as the NFBTE try to balance the intake of apprentices into the industry.

Contracting depends heavily on liaison between the design consultant, quantity surveyor and contractor responsible for a project. To ensure common policies there is a Joint Contracts Tribunal on which the RIBA, NFBTE, RICS and a number of other organizations are represented. This is a key body in relation to the drafting of conditions for contracts, sub-contracts and tendering procedures. Such consultation bodies are fundamental parts of the construction industry, although there are others. They all help to bring together a fragmented industry which, despite criticisms, works and succeeds under many difficult conditions.

The role of the building societies in the UK is an essential part of private sector housing finance. Mortgages depend on deposits and rates offered must compete with other means of saving, including government plans to increase the attractions of government stock, National Savings and other securities.

The UK capital market is highly sophisticated. Public sector housing is government subsidised. Building society mortgages in the private sector attract tax concessions on interest payments. There was also an option mortgage to benefit those who paid little or no tax by enabling them to obtain equivalent financial help by reductions in the rate of interest paid on mortgages, the costs of which were met by the government. There is also a mortgage guarantee scheme which provides for financial advances. This was designed to help those unable to pay initial deposits.

Constructing for public corporations depends upon government finance, but there are partnerships between the public and private sectors on some projects such as oil developments in the North Sea. Many private sector projects, notably large-scale urban developments rely on the expertise of property developers. Specialized development companies can be financed by large insurance companies and pension funds which see investment in property as a protection against inflation. The skills of such developers are recognized throughout the world.

Basis for analysis

What has been explained will, it is hoped, emphasize the complex nature of a construction industry and serve as a basis for the commercial and technical matters which are considered in other chapters. All must be related to the demands of a particular industry sector and to those of international construction for a country or region.

Chapter 2

Industrialization in the UK

Background

One of the features of construction in developing countries is an interest in industrialization. This can be due to the size of demand, lack of management skills, shortage of craftsmen or any one of the many factors dominating attempts to establish an industrial infrastructure. To many, industrialization and system building are synonymous. This is not necessarily true.

Industrialization has been a feature of the UK construction industry for many years, and it certainly has a place in international construction. For this reason, among others, an understanding of what is involved is important. This understanding does not seem readily apparent to many planners who think that industrialization will quickly solve all construction problems, particularly those of developing countries. The concept includes use of systems, particularly those developed by contractors and relying on their special expertise for assembly. It also means greater use of mechanical equipment for site preparation and erection, and use of factory-made components. These have their place in traditional and rationalized traditional building as much as in open and closed systems. Their specification is accompanied by increasing design disciplines and demands for guarantees for performance which influence manufacturing.

A few years ago industrialized building was separate from traditional building, but distinctions are now gone. Public sector planning in the past decades encouraged the use of industrialized building systems for public sector housing. Setbacks in recent years were due to less government emphasis on public sector housing and the encouragement of private sector housing where industrialized building systems were little used. For many years the concept of component building using preferred dimensions and grid layouts, which were given impetus by the design of systems, has been encouraged in the public sector and is influencing private sector housing and all building types.

Optimum output

A problem of using industrialized building systems was to equate optimum output with the size of project in a way which ensured a satisfactory use of capital equipment and continuity of production. Too few sites provided the best conditions.

8

Often the sites for public sector housing projects were for fifty units or less. Few offered the massive scale which led to profitable output and assembly of industrialized building systems and they did not, therefore, produce the revenue expected. The establishment of local authority housing consortia attempted to overcome this, but with less success than was hoped. Largely this was due to the rate of delivery of products which were ordered in bulk, but delivered in a fragmented way. These are few signs that this problem was ever overcome. Large-scale production also requires extensive financing which is usually only easily available to substantial companies. The conditions for optimum output are now frequent in international construction in a way seldom experienced in the UK.

While it must be accepted that UK economic conditions during the past few years have disrupted the ways in which system building progressed, industrialized building systems are successfully used for commercial and industrial buildings or influence their design. Demand for speedy erection to ensure early revenue and reduced manpower on site are key factors in the changing pattern to which component manufacturers are making their contribution. If anything, market factors for greater use of industrialized building components are more evident than ever. System building has made a marked contribution towards improving the standards of housing in the UK. Other improvements are now being made to all types of finishing, landscaping and to the environment overall. Use of concrete components in system building has been widespread, but timber-frame housing and the use of steel-framed systems, particularly for schools and hospitals, is significant. New materials for internal and external use make their contribution to all building programmes, but brick is particularly popular for private sector housing. The construction industry is still largely labour intensive despite the ways in which industrialized building has been introduced. There are, however, shortages of some craftsmen despite relatively high unemployment in the UK.

Innovation

Innovation has been influenced by many UK developments including the preparation of design disciplines and the use of complete systems. Development of materials is under constant review by manufacturers, including new applications of traditional materials and ranges of more recent products such as glass-fibre and plastics. Each contributes to products which incorporate several materials and meet performance requirements. Cladding and internal partitions are examples. Specification by performance is expected to hasten the development of components which meet exacting user tastes as technology advances, including thermal and acoustic qualities, visual appeal and cost effectiveness.

Dimensional coordination

Apart from the influence which earlier industrialization had on the design of buildings and, indeed, the structure of the market, it also showed needs for dimensional coordination. Much work has been done by all sectors of the industry to produce comprehensive dimensional design disciplines, partly as a result of changing to the metric system.

One of the first recommendations for the coordination of dimensions in building to be published in the UK was the BS 4011 : 1966. This set out basic sizes for building components and assemblies and was intended to encourage the wider use of factory-made components and assemblies and a reduction of site work and ranges of sizes. The standard set out recommended basic sizes based on multiples of 300 mm and 100 mm, and 50 mm and 25 mm for measurements up to 300 mm.

At the start of the programme for the change to the metric system in the UK, it was necessary to determine a dimensional framework from which basic sizes of dimensionally coordinated components could be derived. Several UK government departments concerned with public sector building programmes prepared extensive recommendations for use in the public sector, and were given to the British Standards Institution (BSI) which was responsible, through a committee structure composed of representatives from the public sector, professional and trade organizations and industry generally, for planning the design disciplines explained. Eventually BS 4330 : 1968, which set out recommended controlling horizontal and vertical dimensions, was published. It gave guidance on design sizes for floor to floor and floor to roof heights; floor to ceiling heights; changes in level; horizontal spacing between loadbearing walls and columns; zones for floors, roofs, loadbearing walls and columns and heights for door and window heads and sills. It provided for the horizontal spacing of dimensions between axes or between boundaries of zones for loadbearing walls or columns and their layout in relation to grids and was a logical follow-up to BS 4011. The key recommendations of BS 4330 and definitions used are:

Floor to ceiling heights

Heights in multiples of:	300 mm	100 mm
	1 500*	
	1 800*	
	2 100†	
		2 300
	2 400	
		2 500
		2 600
	2 700	
		2 800
		2 900
	3 000	

* Applies only to farm buildings   † Applies only to domestic and lock-up garages, multi-storey car parks and farm buildings.

Greater heights in multiples of 300 mm from 3 000 mm to 6 600 mm and thereafter in multiples of 600 mm. In addition to the values in the table, 2 350 mm
may be used, for housing only, in conjunction with a floor to floor height of
2 600 mm.

Heights of zones for floors and roofs

Heights in multiples of:	*300 mm*	*100 mm*	*50 mm*
		100	
		200	
			250*
	300		
		400	
		500	
	600		
	900		
	1 200		
	1 500		
	1 800		
	2 100		

* Applies only to housing for use in conjunction with the floor to floor height of
2 600 mm and floor to ceiling height of 2 350 mm.

Greater heights in multiples of 300 mm from 2 100 mm.

Widths of zones of columns and loadbearing walls

Widths in multiples of:	*300 mm*	*100 mm*
		100
		200
	300	
		400
		500
	600	

If greater widths are required, they should be in multiples of 300 mm as first preference or of 100 mm as second preference, in accordance with BS 4011.

11

Changes in Level

Range (mm)	Heights in multiples of (mm)
From 300 to 2 400	300
Above 2 400	600

Changes in level of 1 300 mm, 1 400 mm, 1 700 mm, 2 000 mm and 2 300 mm may be used for housing in conjunction with a floor to floor height of 2 600 mm.

Floor to floor and floor to roof heights

Heights in multiples of:	300 mm	100 mm
		2 600*
	2 700	

* Applies only to public sector housing.

Greater heights in multiples of 300 mm from 2 700 mm to 8 400 mm and thereafter in multiples of 600 mm.

Spacing of zones for columns and loadbearing walls

Range (mm)	Spacings in multiples of (mm)
From 900	300

800 mm may be used for housing only.

Controlling dimension

A dimension between key reference planes such as a floor to floor height. Controlling dimensions provide a framework, within which buildings may be designed and to which building components and assemblies may be related. Intermediate controlling dimensions are sub-divisions of the main controlling dimension framework.

Controlling line

A line representing a key reference plane. Controlling lines for vertical dimensions represent boundaries of zones for floors and roofs. Those for horizontal dimensions represent the boundaries or the axes of loadbearing walls and columns or the boundaries of zones. Axes do not necessarily coincide with centre lines.

Zone

A space between vertical or horizontal reference planes which is provided for a building component or group of building components which do not necessarily fill the space. Zones for floors and roofs contain the structure and may also include finishes, services, suspended ceilings and similar components and, where appropriate, allowances for camber and deflection. Those for loadbearing walls and columns contained in the structure may also include finishes and casings. Provided the use of coordinated components is not inhibited, a building component or group of building components may extend beyond the boundaries of the zone, and finishes may be placed outside the zone boundaries.

Controlling zone

A zone whose size is in accordance with BS 4330.

Neutral zone

A zone whose size is not in accordance with BS 4330, but whose size should preferably be in accordance with BS 4011.

Apart from the advice which is set out in BS 4330, additional BSI information explained ways in which dimensional recommendations were developed, particularly the arrangement of building components and products in the six functional groups. These are: structure; external envelope; internal sub-division; services and drainage; fixtures, furniture and equipment and external works.

The relative importance of certain components for dimensional coordination was established. There were some components for which dimensional coordination was essential. Others depended upon this, but others were relatively unimportant. Products specified in the functional groups structure for the external envelope and internal sub-division were given ranges of basic spaces.

International System Units were published in BS 3763 : 1964 as a part of the change to the metric system, and are relevant to international design. This is only one of many supporting documents. At an early stage it was evident that a glossary of terms was essential. An earlier standard was revised and BS 2900 : 1970 was published. Other supporting design documents cover accuracy in building, tolerances and fits, combinations of sizes and tolerances and fits for building.

Design disciplines introduced in the UK have been the work of cooperation between government departments and all sectors of the construction industry. Impetus, however, depends to a large extent on market factors. Absorbing the design recommendations described places demands on industry skills. Comprehensive applications of dimensional coordination are likely to be long-term. Nevertheless, the work done has led to a comprehensive set of disciplines which can have a profound influence as building technology changes and is increasingly studied and used throughout the world.

Performance specification and performance standards

Use of performance specifications, largely by government departments, was a major factor which led to studies by BSI of whether or not performance standards should be introduced. Performance specifications issued by government departments and by local authority consortia covered, among other things, cladding panels, doorsets, partitions, roof-lights and windows. One of the major factors considered when using performance specifications was the point at which they should start. In the UK early discussions about this were related to levels of construction and manufactured products. The heirarchy used was:

- Construction undertaking: Scope of performance specification or standard.
- Major conurbation: Broad social requirements.
- Small town or zone: Functional requirements for self-contained areas.
- Individual building: Functional requirements for internal areas.
- Individual room or space: Functional requirements for specific areas.
- Civil engineering works: Functional, manufacturing and constructional requirements expressed in codes and standards.
- Building complexes: Functional, manufacturing and constructional requirements expressed in codes and standards.
- Buildings: Functional, manufacturing and constructional requirements expressed in codes and standards.
- Products: Scope of performance specification or standard.
- Building elements: Functional, manufacturing and constructional requirements.
- Assemblies of components: Functional, manufacturing and constructional requirements.
- Components: Functional, manufacturing and constructional requirements.
- Formed materials: Quality, manufacturing and assembly requirements.
- Unformed materials: Quality and manufacturing requirements.

Performance standards for some materials and products have existed as British Standards for a long time and extension to components, assemblies of components and elements appears reasonable, but application to complete buildings and towns is less practicable. All designs, irrespective of product, require that a specification be given. However, any specification is only the sum of what is possible through the use of individual components and it is at this point that the concept of performance specifications and standards in the construction industry appears most relevant. Those responsible for initiating performance specifications saw them as a method to obtain the best products at economic prices, but time will be needed to determine success achieved.

International liaison

Work on standards being done in the UK is not separated from that being done elsewhere. It also contributes to the planning of others. Much of that relating to dimensional coordination and specification by performance has been referred by

BSI to the Comité Européen de Normalisation (CEN) and to the International Organisation for Standardisation (ISO). Clearly the advanced industrial countries, particularly those of Western Europe, are likely to assess the potential which exists in the use of concepts outlined for the short and medium term. They are further steps towards a greater industrialization of construction, but much depends on the structures of individual construction industries and the application of factory-made components within industrialized building systems, rationalized traditional or traditional building. Continued developments in the UK depend on what is technically and commercially realistic in relation to current or future construction markets at home and overseas.

Future contributions

Design and other disciplines planned and influencing construction are only one part of improving industrialization and building technology, output and user benefits. Economic factors are equally important. No planning can be successful without forward thinking which is equally essential in a developing country as in a highly industrialized country such as the UK. There is no single key factor which is dominant. Information, optimum production and delivery rates and a sound economy are linked. Research and harmonized national and international legislation, regulations and standards are key factors which must, in turn, be related to the needs of individuals and to commercial realities.

Success of design and other disciplines depends considerably on users and their professional designers. Despite the size and influence of many manufacturers, the extent to which they can dominate the market is limited. Generally it is the capital intensive sector, which produces such items as sheet materials, which has the greatest influence. Most components are made by many smaller companies, who are subject to market pressures and user and designer demands. Planners and designers, therefore, must understand how manufacturers and contractors work and their part in the overall construction industry structure. These are remarkably similar in all countries and must be related to central and local government planning, finance available and, above all, industry skills and capacity.

Chapter 3

Overseas markets. Marketing, appraisal, decision-taking and planning

Principles

Success in international construction depends partly on marketing. Principles of this are similar for consultants, manufacturers and contractors, but manufacturers generally developed their methods earlier and, therefore, often have greater experience. The techniques used have grown in detail and sophistication during the past two decades. In many industrial countries, the essential demand for all types of construction has now been met. This does not mean that increased standards or higher output is not needed, but it does mean that demand is more selective. In recent years a buyers' market has developed and all in the construction industry must market their skills and products. Consultants, subcontractors and main contractors are no exception and find it necessary to adopt a marketing approach. Those responsible must assess social and political trends, population movement, growth of development areas and construction needs in ways not previously necessary. The methods used are as relevant to international markets as to those at home.

Definitions

Despite jargon, marketing is essentially assessment and meeting a demand for products and services, and ways to ensure optimum use of existing capital equipment and skills. Without appraisal of these primary objectives, marketing could become a meaningless word and lead to new ways of describing the process of entering and staying in business, which is now largely the responsibility of those concerned with marketing who have, to a large extent, replaced earlier owner-managers. A few decades ago, limited technical skills, concentration of purchasing power and transport difficulties were key regulators of demand and production. They no longer apply. Advancing technology, rising discretionary incomes and easier communications give rise to new market conditions. These, plus social pressures, mean different market regulators. The capital investment required for the simplest production can now be so vast that it is no longer sensible or possible to enter any market without considerable market research. Changing factors of production can require manufacturers to operate in larger units, for which corporate planning and needs for increased production to

16

achieve economies of scale may mean higher capital investment. The construction industry is as influenced by these trends as other industries.

Scale

Scale need not be dominant, but the complex structure of a construction industry prevents the type of function undertaken by the market-stall owner and salesman. He operates a marketing role in model way if he is to remain in business; correct assessment of market prices and customer requirements, flexible buying and pricing policy based on time of day or week, and a ready price adjustment to match competitors' and cash sales. This type of business usually requires cash purchases before sale. Most construction industry products are still made ahead of orders and the role of marketing must be to ensure that sales match production and vice versa. Cheap and simple products are affected by this as they may involve high capital investment in a similar way to the construction of a major airport or item of capital equipment. Neither is usually constructed without a planned use and order

Speculative construction

What has been said so far gives a clue to the success of some of the speculative housebuilders in the UK. A very high percentage of houses is built by many medium and small-size companies who have far greater flexibility and less overheads and can, therefore, adjust to market conditions more quickly than their larger competitors. They may not, however, usually be able to match the financial and management resources of their larger competitors. For the supplier of products who may be second or third in a sequence of planning a building, overhead costs are equally important. Market analysis is a vital part of a manufacturer's work and may not be influenced by one market only. Sales may be made to a variety of industries; motor, shipbuilding, furniture and others. While this creates problems, it can help to prevent violent changes of output as sales in different industries fluctuate. Overseas markets are equally essential.

Objective

One of the most vital functions of anyone responsible for marketing is the establishment of an objective. An assessment of this will influence raising finance, profit targets and methods of achieving them, location of manufacturing and distribution resources and, above all, it will encourage a definite purpose and avoidance of activities likely to dissipate resources. The latter may, initially, appear to be a diversification of interests, but, unless they are part of a plan to avoid fluctuating profits and use of output, diversification can also lead to various management functions and organizational difficulties.

Contractors and manufacturers are usually the only members of the industry dependent upon expensive capital equipment, although some contractors operate with little capital equipment. The net trading profit on capital employed can fluctuate considerably and many may operate with a very low margin of profits. As a result financing costs and short- and long-term debts demand a continual emphasis on the profitability of all undertakings and which influence objectives most strongly.

Market assessment

Assessing a potential market and the ways in which meeting it could influence production is a massive task. It is made difficult by the ways in which all aspects of national policy influence construction in the public and private sectors and as a result have a bearing on industry policies. Each development must be analysed. Those that may have only relatively limited application for a sector of the construction industry must be fully studied as it is important that any planning undertaken in the industry should lead to compatible activities

Production

Production may be labour or capital intensive. For many manufacturers expensive plant is controlled by relatively little labour. Construction of such a plant may depend on vast manpower and is labour intensive. The manufacture of construction products is increasingly capital intensive and market assessments to ensure correct investment in production capacity are essential to achieve satisfactory profit levels. Manufacture of many components continues to be labour intensive, although the manufacture of such items as timber doors, metal windows and many fittings is becoming more capital intensive. Rising manufacturing and labour costs may lead to changes, but capital costs could result and go beyond the scope of many small companies. The result will be a greater demand for marketing skills which could also be outside the resources of many companies.

There is little design, manufacture or construction beyond the skill of European industry, provided finance, equipment and labour are available. Despite the usual problems of obtaining these, which are subjects for early market study and influence location of activities, they are separate from market studies to establish the likely success of a venture which may have national and international restraints. A primary function of a marketing study or strategy must be to prove that a service is needed or to explain a need which can be filled. Such functions may be supported, after research, by advertising and technical sales literature. These are, however, only a start to obtaining an order which will utilize output capacity as much as possible and maximize profits. All other functions – assessing trends, research and development and other facets of industrial administration – must be directed towards these objectives.

Marketing support

Consultants pursue a marketing role when creating a creating a design or assessing the viability of a project in the same way as others. At a more advanced stage, designs may improve layout or circulation, increase plot ratios or densities and, therefore, rents. Selection of products and materials to achieve the right end product at the best price are also marketing functions. They do not end when a design is completed. A service, once selected, probably as a result of advertising, direct selling and user and manufacturer negotiations, may be forgotten when it is included in a specification. The total marketing function has not been achieved until a service has been completed and payment received.

Production and sales are usually supported by sales service organizations which work closely with those responsible for market analysis and planning, production and quality control. Such machinery exists to provide a service and ensure sales and goodwill for future sales, without which, and the resulting satisfied customers, no amount of market analysis and prediction of trends would have any meaning whatsoever.

Ways in which market factors of the type outlined are assessed differ within the many organizations which make up the construction industry in the UK, as in other industrial countries, but whatever the method or information used, systematic analysis and the creation of a plan based on sound judgement must be the aim. Organization of the approach needed is essential. This is as relevant to deciding a design, for the production of a small number of doors or windows, large-scale output of a particular material or construction of a house or major project. Too often the basic techniques of analysis are wrongly assumed to be known, or are ignored.

Market domination

Despite the size and influence of many companies, the extent to which they can dominate a market is limited and generally it is the capital intensive sector which has the greatest influence, but only in their own countries. World competition reduces power in international markets. Companies are subject to market pressures and design and user demands throughout. The law of diminishing returns constantly influences a manufacturer. This can be as important to designers, to whom a small addition of staff, office space and equipment may be a large percentage of overheads, and may not be justified in relation to risk or potential returns. For even the smallest manufacturer, the way in which diminishing returns influences profits may have to be related to very considerable investment. For large companies, millions of pounds may be involved. Commitment to the market is, therefore, undertaken only after most serious study of the economic laws and factors of production which govern marketing and all aspects of supplying products.

It may be argued that a reserve of mobile and uncommitted manpower is essential for balanced marketing. No one would deny that some reserve of cash, either as cash flow or credit is essential in any undertaking. Some reserve of housing and, to a lesser extent, industrial and other buildings, are also necessary. Lack of surplus accommodation at a reasonable price discourages mobility as much as other social factors. Marketing in the construction industry helps to achieve a balance.

Basic data

There are many factors which influence marketing decisions. Definitions of basic data necessary for these may vary. Some may be factual and scientific. Others will depend on the experience, shrewdness and analytical ability of an individual or collection of individuals. Marketing within a construction industry is the key to matching technical developments, customer requirements and commercial realities. Good organization is an essential key to good marketing in any country.

Overseas marketing

Establishing an overseas market need not differ from procedures used in the UK. A decision is required on the reason for pursuing overseas work. It may be because UK turnover is expected to fall, or because diversification into a foreign market is considered worth while. A primary reason may be to use unfilled capacity, to avoid redundancies, or because overseas profits could be better or have taxation advantages. Each factor will influence decisions which have to be made. They should be based on more than short-term involvement in a particular market.

No company should pursue a market overseas which is different from that operated in the UK. The international demand for construction is so extensive that it could encourage entry into markets which are unfamiliar, but generally it seems best to pursue a market which is known best. This may determine the selection of a country or area likely to have the best potential for a particular service.

Method of working

In most cases it will be necessary to establish an overseas company. Depending on local law, this could be a fully-owned subsidiary, but it is more likely to be established with local partners. Local law may not permit the establishment of a corporate body, and it will be necessary to form a partnership. The existing company structure, financial organization and taxation may encourage the establishment of an offshore holding company. Management from a UK base may be possible, particularly in the case of manufacturers, but for contractors a local presence will be essential. A local agent may be necessary to support selling and management, and could be an essential part of establishing a local organization and obtaining initial contracts. Preliminary decisions on the policies affected by these factors can be taken at an early stage.

Selection

It is worth while considering whether or not an objective is the public sector market, the private sector market or for markets in both. Each has advantages and disadvantages. The public sector market can be more secure financially, but demanding bid bonds, performance bonds and contractual requirements could be discouraging. It may be decided that private sector markets, where decision-taking may be quicker, provide a more satisfactory basis for working. This market may be the only one for certain types of construction. It is also a market which can depend on a local partnership and finance.

In many developing countries there is no shortage of finance. However, public sector experience in some countries in recent years has encouraged those responsible for signing contracts to be very cautious. Local finance may be readily available for private sector work, but there is a tendency for local partners to require the financial involvement of foreign partners. Earlier there was great emphasis placed on obtaining commercial, technical and management skills, but finance is now sometimes required as evidence of continued involvement in the market concerned.

To assist making the type of decisions outlined, information can be obtained from a number of government, industry, and commercial and other organizations. Each has valuable information which can be obtained before spending money on overseas visits.

When establishing a marketing plan the key functions are to identify and assess demand, create demand and sales and to meet and service the demand established. Fundamental economic and construction industry factors which must be considered are: demand; output; finance; prices; costs; manpower; and delivery. Steps to be taken to prepare a marketing plan can be equally relevant to consultants, contractors, manufacturers and others in the construction industry, and include study of such factors as:
Structure of the market

- Specifiers. Differences between roles of architects and other designers and consultants.
- Contractors. Contractual conditions, differences and competition.

- Purchasing methods.
- Distribution patterns.
- Transport, port and airport facilities, and customs clearances.
- Supply of plant, materials, products, competition and tariffs, and other trade barriers.
- Professional, trade and employers' associations and trade unions. Role and significance and employment of manpower.
- Land availability, costs and ownership.

Specification
- Legislation affecting construction and use of products.
- Standards required.
- Assembly responsibilities.
- Testing of products.

Legal
- Liabilities.
- Insurance requirements.
- Rules governing partnerships and companies, and use of agents.
- Labour laws.

Finance
- Sources.
- Costs.
- Taxation, repatriation of capital and dividends.
- Bid or performance bonds.
- Payment terms.

Climate, employment, local customs and religions
Security
Communications

Principles which apply in the UK can apply equally well overseas, but a decision on how a market can be sustained is important. It may be possible to sustain markets in the UK or Europe, but to support operations in a developing country in the Middle East, Africa, or South America is different. The decision, therefore, is whether there is the management and will to service a demand.

A plan for developing a market depends on a concept, the product or service proposed and definitions of the financial, commercial, management, technical and legal constraints. These may dictate a feasibility study. Such a study should define the objective to be achieved in relation to the product, service or expertise offered and market research to be undertaken. Much may be done in the UK. Having defined an objective, it is essential to assess the size of market, buying pattern, methods of distribution, output required, pricing structure, competition and facilities required to support an existing or a new demand. A likely sequence of work for a marketing plan, business development and a feasibility study are set out below.

Marketing plan

Function or objective	*Method*	*Information required*
Identify and assess demand.	Assess market trends. Locate demand. Analyse business needed.	Public and private sector expenditure on various construction types or sector. Economic forecasts. Availability of finance, manpower and communications. Customer reactions to existing and proposed business.
Create demand and sales.	Create business policy, including pricing and credit structure. Establish sales and distribution network and promotional support such as public relations activities.	Profits and sales targets. Methods of selling. Advertising programme and methods. Public relations objectives and standards.
Meet and service demand.	Control of production and quality. Budgetary control. After-sales service. Business improvement and innovation.	Likely pattern of orders. Target output. Business duration before changes. Replacement and support needed for a demand pattern.

Business development

- Concept.
- Proposal.
- Define financial, commercial, technical and legal constraints and standards to be achieved.
- Feasibility study.
- Assess feasibility study.
- Preliminary marketing planning.
- Design.
- Assess business in relation to feasibility study and market plan.
- Production tests.
- Final marketing planning.
- Launch.
- Assess degree of success of launch.
- Assess need for marketing plan changes.

Feasibility study

- Objective of a study.
- Define business or likely market.
- Initiate market research to determine need for business or likely market.
- Assess:
    Standards required;
    Size of market;
    Buying pattern;
    Method of distribution;
    Output required;
    Pricing structure;
    Competition;
    Relationship with existing business within own company or within a group policy.
- Research and development required.
- Facilities existing and required.
- Costs and financing.
- Sales factors.
- Design factors.

Market assessment costs

Finance required for a construction markets assessment could be extensive. For major marketing functions and pursuit of major contracts very large amounts could be necessary, and even then a signed contract may not be obtained.

Statistics

Statistics can be essential. Economic forecasts, expenditure planned within public and private sectors and expenditure on specific building types by regions can be particularly helpful. Population distribution and likely growth, communications, gross national product forecasts and contributions by particular industries may also indicate patterns of demand. More important, however, is the specific information which will assist the pricing of a contract or tender. Cost and production indices for developing countries may not be available, but some planned expenditure may be, although not related to resources. It is essential to know the availability and distribution of manpower, structure of the local construction industry, including main contractors and sub-contractors, and their financing methods, ways in which orders are placed and the output of materials and products. It is particularly difficult to obtain up-to-date prices which operate in most overseas markets. For this reason alone, a local agent or office may be essential. Suggested essential statistics required for a construction industry marketing plan are:

Preliminary planning

- Economic forecasts.
- Planned expenditure within public and private sectors.
- Expenditure on specific construction types by regions or planning and development areas:
    Housing;
    Schools;
    Hospitals;
    Industrial buildings;
    Offices;
    Roads;
    Special projects and civil engineering programmes.
- Population distribution and likely growth.
- Communications.
- Gross national product forecasts and contributions by industries.

Detailed planning

- Cost and production indices.
- Gross domestic product related to construction.
- Patterns of output and new orders.
- Manpower. Availability and distribution.
- Structure of industry. Type, size and distribution of main contractors, specialist contractors and sub-contractors.
- Financing methods. Sources of finance from public and private sectors.
- Output of materials and products.
- Output of construction by types.
- Size of competitors' markets and scale of operations.

Restraints

Characteristics of climate, transport, distribution, communication, supply of products and manpower and technical limitations create their own difficulties in many developing countries, but major problems appear to be financial and contractual. On entering a market it will be essential to decide at an early stage whether the financial demands placed on a company by the need for bid bonds, performance bonds, insurance against unfair calling of bonds and uncertain payment are acceptable.

Always assess the likely trading structure. Apart from selecting the best overseas corporate structure, it will be necessary to decide how far existing legal services are able to cope with the demands of international markets. It will also be necessary to decide whether financial services are adequate. Contractors must ensure familiarity with the type of contracts used overseas, the Conditions of

25

Contract (International) for Works of Civil Engineering Construction and for electrical and mechanical works usually known as FIDIC, ways in which other forms of contract can be modified, conditions likely when contracts are financed by the European Development Fund and ways in which they relate to local laws and needs. In turn, it is necessary to consider whether an estimating department will be able to meet the demands of overseas work or whether those concerned would find it impossible to obtain prices which would enable the preparation of a satisfactory bid. Any decisions to establish a local company must in part be influenced by decisions on recruiting management and manpower locally or from the UK. A new approach to recruitment may be necessary and will be influenced by salaries, pension rights, insurance, needs of families and many other factors, all of which will test an organization to the full.

Agents

If an agent is employed an early assessment of how he is to be paid must be made. This will be affected by decisions on methods of working and a contract for employment, which may be for one or more years to an exclusive contract. It is reasonable to expect that an agent will advise on pricing factors, availability of labour, plant and building materials and products, shipping and distribution factors, demands likely to arise on mobilization and on factors affecting the housing of personnel. An agent can be helpful when preparing design submissions, prototypes and bids and on competition. Advice on methods of payment which may be expected should also be given.

Decision-taking

There are many planning factors to be considered in relation to an overseas market. At all stages it will be necessary to make decisions and probably form a plan. The following is a suggested procedure for analysing a business or marketing problem.

Analysis framework

- Statement of objective or task.
- Analysis of factors relevant to the problem.
- Each separate factor should lead to a separate conclusion.
- Summary of conclusions.
- Reject conclusions that are irrelevant or of lesser importance.
- Select course of action based on priority conclusions.
- Prepare plan or decision.

Plan or decision execution framework

- Statement of background information.
- Objective or task.
- Execution. Relate each requirement to key organisation functions and those responsible:
  Organization structure
  Organization policy
  Structure:
    Marketing;
    Production;
    Sales;
    Finance;
    Manpower;
    Research or development;
    Administrative, finance, legal, advisory and other services.
- Communications:
  Internal;
  External.

Demands on management

Throughout there will be questions about the difficulties and financial demands which will be placed on a company through overseas operations. In many overseas markets, construction costs are two to three times higher than for similar activities in the UK, and place demands on resources so that one or two contracts only stretch existing management and financial resources to a point that is unacceptable. In all cases it is necessary to obtain as much information as possible to prepare a plan in the UK for a subsequent visit to a country or area, to ensure that the management structure, finance and will exist.

Chapter 4

Site reconnaissance and tendering

Importance of site visits

Marketing appraisal and decision-taking are functions undertaken by all in the construction industry, and, indeed, other industries, although many aspects of marketing may relate more to manufacturing or providing manufactured goods. Site reconnaissance and tendering are, however, well-defined functions of a contractor. Even so principles governing preparation of a tender can be adopted for other purposes, and certainly most market factors will influence tendering.

At an early stage a fact-finding visit to the country and site concerned will be necessary. This will also assist making valuable contacts in the area concerned. The time available to make enquiries in any depth may be limited, and it will be of great help if information is also obtained before leaving the UK.

Organization of the construction industry

Understanding the local organization of construction, and such things as the size of contracts, financial strength of contractors, suppliers of products and the importance of sub-contractors must be a basis for preparing any tender. It is also particularly important to understand the powers of the architect, consulting engineers, or other professions when a bid is selected. They may be nationals of the country concerned, but are more likely to be from a variety of other countries. Methods of cost control will depend largely on their origins and training, as will control of a project and methods of approval at the various stages of design, tender and final approval. Often the availability of trained designers, technicians, craftsmen and unskilled labour will have influenced their planning, and need to be clearly understood before tendering.

A tender should indicate whether or not the customer is a municipality, state, or the central government. This will control the customer organization for placing contracts, including the basis of public accountability and the relative importance of officials and elected representatives. To understand this fully, statistics showing output may be required, preferably with statistics of planned output.

Analysing a tender

Whatever the experience of a company when tendering for an overseas contract a first essential is to assess the status of a potential customer, particularly in relation to public and private sources of payment. Often pre-qualification will be necessary, which can be time consuming and costly, and viability should be related to others known to be bidding. Too often those responsible for awarding international contracts have allowed many to pre-qualify, and have created abortive work as a result leading to unrealistic tender prices. The lowest bid may not always be accepted for a variety of reasons which can usually only be assessed locally.

It may be necessary to bid with a local partner. This should only be done after the most careful study of the legal, financial and taxation restrictions in the country concerned. There may be local contract laws or the Conditions of Contract (International) for Works of Civil Engineering Construction may be acceptable. There are also similar international conditions for electrical and mechanical works. Throughout, the language and law applying to the contract are of paramount importance. Bids may also be governed by bid and performance bonds. Factors governing an assessment of the customer, pre-qualification, contract conditions and bonding include:

Customer

Employer; consultants, architects, engineers, quantity surveyors and others involved and economic and political conditions.

Description

Location; estimated cost; commencement and completion dates; penalties for delay; maintenance period; payment terms and bonds required; contract law; design drawings and measured bills of quantities and nominated subcontractors.

Designs

Changes likely; undefined work; sequence of work and extra work; inspections necessary or desirable and policy on any expansions of the contract.

Finance

Currency, rates of exchange and restrictions, borrowing costs; company taxation, stamp duty and tax; employees' taxation; cost of living and cost of construction indices.

A site survey, economic and political assessments and manpower study are essential parts of any agreement to bid by a company. They may be undertaken before or during analysing the matters already described, but should certainly not be delayed until afterwards. Much of the process of decision-taking is dependent on many inter-related factors and decisions. The availability of manpower, its skills and politics will clearly govern early contract decisions and the technical capacity to achieve success. Essential matters for study and decision which may relate to local employees, foreign manpower and, of course, to UK expatriates are:

Manpower

Labour laws, restrictions and permits; wages rates; bonus incentives; compensation, termination pay and insurance; hours worked daily and weekly, overtime rates and holidays; availability of managers and supervisors, costs and conditions of employment; availability of manpower by trades and costs; subsistence payments and local allowances, including medical treatment; union organizations; labour and staff cost inflation and comparative efficiency ratings of local and UK projects.

Manpower support

Accommodation costs and manpower needed, and need for special provisions; camps and offices; water supply; food; sanitation; hospitals and health needs; schools; fire and police services; recreation, travel and communications; laundry facilities and special clothing provisions and security.

Site surveys

Many of the commercial decisions governing a tender may be taken in the relatively easy conditions of an office or boardroom, and away from the country concerned. Those responsible for technical decisions have no such advantages. Only demanding, time consuming and expensive site surveys can provide the information necessary. All will determine the success and profitability of a project, and will have a bearing on the legal, financial and similar matters which could disrupt completion and payment for work done. No matter how detailed a design or comprehensive the specifications they will be irrelevant if physical factors make work impossible or impracticable. Climate, land, soil and rock formation, water and access to a site must be studied. So, too, must the availability of plant, transport, energy, the restrictions imposed by local controls and, above all, sub-contractors and their resources. Matters which must be studied include:

Access

Obstacles; rivers and crossings; transport restrictions including need for temporary roads or bridges; maps available and security.

Land conditions

Topography, terrain, vegetation and drainage.

Geology

Top and sub-soil, soil movement and erosion; physical characteristics including faults, rock, ground water and springs.

Surface and ground water

Flow and floods; predictions; tide tables and equipment needed.

Climate

Seasons; temperatures, impact of rainfall, mud, floods, snow and ice, hurricanes and dust; earthquakes and fire hazards

Property

Boundaries and easements; access and adjacent owners; purchase and rental requirements; riparian, mineral, timber and dumping rights and compensation.

Energy

Sources and characteristics including type, capacity, transmission and storage and costs.

Sub-contracting resources

Local sub-contractors; labour, haulage and other rates; existing work in progress; main contractors working in the area and competition and interest in other projects.

Local controls

Local authorities; bye-laws and permits required for contracting operations and work and other permits needed.

Tools, plant and transport

Availability and cost of tools, plant and transport; servicing facilities; siting of workshops and hard-standings for plant and transport and scaffolding availability and costs.

31

Communications

Telephones and radio; air services and costs; rail, road, river and canals capacity and costs and efficiency ratings and bridges and capacities.

A further technical study which will influence a bid is the availability and cost of materials, notably local supply and manufacturing facilities, merchants, the procurement organization to be established and the freight facilities available. Some purchases may be undertaken from the UK, but local needs and supervision are paramount. A careful understanding of local conditions including the strength of an agent will be needed. Freighting factors will also need early study and on assessment of shipping and airports available; charges; lightering and stevedoring; import and export duties and wharfage, unloading facilities and storage.

Housing

Case study

In many overseas countries housing demand is high and is an essential part of any construction programme. Often tenders are invited for small and large projects on the most limited information. An analysis of meeting such a tender offers a very useful case study of information needed for most types of tenders.

Preliminary assessment

Forecasts of the proportion of dwellings to be erected under various density headings will be needed. These could be divided into low, medium and high ratios. Low density is the type of development most likely in rural or small urban communities, and in UK terms might be up to fifty dwellings per hectare. Medium density in UK terms would be from fifty to one hundred dwellings per hectare and could be expected to be found in all types or urban communities where land supply is not too limited. High density over a hundred dwellings per hectare may be expected to be found in areas where land is short or expensive. Land costs and population estimates must support this information with knowledge of the acceptability and unacceptability of dwelling types. As an example, establish the acceptability of terraced houses compared with single-storey dwellings, and maisonettes, and the optimum size of family unit to be housed in a high-rise block, if these are acceptable.

Much will depend on present and projected space standards. For public sector housing there may already be space standards. A dwelling size may be expressed either in the number of persons or the number of habitable rooms, or the overall dwelling area in square metres. The information should show the division and mix of dwelling sizes.

Development of dimensional planning and coordination could be important, but may have had limited progress. However, information about the climate and requirements for sunlight or shade, sound-insulation, through ventilation, internal kitchens and bathrooms and regulations which will affect the layout of dwellings are essential. There may be unusual requirements influencing structures due to the probability of earthquakes, fires, typhoons and similar risks which will be critical to preparing an acceptable design or tender.

The size of a project, space and other standards may dictate the type of design or materials to be used. New materials may be acceptable for items such as floor coverings, partitions, doors, wall coverings, as well as for the more functional ones such as baths, showers, toilets and kitchens. Special design factors, particularly for baths, heating and kitchens must be understood. The external appearance of multi-storey housing may differ from traditional designs, which governs the use of external materials. Acceptability of materials such as aluminium, asbestos and plastic sheeting, bricks, plain concrete, exposed aggregate concrete, mosaic tiles and curtain walling must be determined. In modern high-density living greater attention is being given to the landscaping and to the provision of reasonable amenities for the community. This could lead to a need to provide such facilities as sports fields, courts, gymnasiums and swimming pools for the large high-density communities. Provision for access, parking and storage for motor cars is an important part of modern community planning. Forcasts of ownership of cars per household will influence this planning. All may, nevertheless, be subject to the special restrictions of local climatic, political and, above all, religious factors.

Housing programmes may be a part of developing an infrastructure. They may also dictate needs for this. Educational buildings may be required, including training colleges and universities. In some cases these may form a part of a large housing community. Other community buildings, those for commercial and industrial use and transport and public utilities may be planned as a part of a housing development.

Aide-mémoire

A summary of factors on which information is needed for housing tenders is as follows:

Organization of the construction industry
- Key professional and trade associations.
- Architects, consulting engineers and other professional advisers and their roles.
- Turnover, asset value and capital structure of local contractors.
- Values of contracts.
- Use sub-contractors.
- Contractors' buying methods.
- Builders' merchants distribution network.
- Size of suppliers of materials and products.
- Turnover, asset value and capital structure of local manufacturers.

- Contracts approval.
- Methods of design control, tendering and selection of products.
- Availability of qualified designers, technicians, craftsmen and unskilled labour.

Organization of customers
- Relative values of public sector and private sector work.
- Output and output planned.
- Regional planning.
- Design and placing of contracts by local authorities.
- Local authorities housing, hospitals and schools programmes.
- Housing cooperatives.
- Central government subsidies.
- Methods of financing developments in the public sector.
- Private sector financing methods.
- Development companies operating in the private sector.
- Insurance companies operating in the development of buildings.

Planning requirements and standards
- Existing and future planning density standards.
- High-, medium- and low-rise developments.
- Land costs in urban and rural areas.
- Planning or social resistance to particular building types such as high-rise or terraced developments.
- Space standards in towns and elsewhere.
- Cost yardsticks.
- Size of families occupying high-rise dwellings.
- Special requirements for sound-insulation, sunlight, ventilation, kitchen and bathroom layouts.
- Sound-insulation.
- Standard of finishes.
- Important structural or other requirements resulting from earthquakes, typhoons and fire risks.
- Dimensional coordination.

Sizes and types of housing contracts
- Size of contracts in the public and private sectors.
- Work expected to be undertaken on new towns, urban renewal and slum clearance.
- Main areas of redevelopment.
- Preferences for particular materials.
- Materials in short supply.
- Production capacity and manpower for making materials and products.
- Preferences for certain types of external finishings.
- Preferences for certain types of structures.

- Planning of amenities.
- Factors influencing the design of baths, heating and kitchens.

Other buildings
- Programmes and plans for building schools, hospitals, supermarkets, community meeting places and transport facilities, and other public utilities and works.
- Factors influencing the design of schools, hospitals and shops.

Success depends on realism

Throughout the assessment of a tender there will be similarities with UK practice, but the governing factors will be those of climate, manpower and materials available, and communications. However, success or failure will too often depend on additional local political and economic restrictions which are not based on objective judgement of the type normally associated with UK practice. The emphasis of the bid and eventual project is, therefore, not so much on what is required as on what is acceptable, permissible and possible.

Chapter 5

Relationships between main contractors, specialist engineering contractors and sub-contractors

A most important subject

It would be all too easy to start writing about each subject related to international construction by saying that it is one of the most important. Indeed, the complexity of any construction project is such that no part can be studied in isolation, nor be truly considered as less important than any other. Design will influence assembly, selection of products will be governed by specifications, and methods of working may be dictated by contractual, financial and other constraints. However, construction is essentially the assembly of products by people, usually under conditions which at best can be difficult, hazardous and subject to the domination of the weather and environment. It is seldom carried out in conditions which can match those of even the most rigorous of factory working. The final product, whether a building or civil engineering undertaking, on which will depend the provision of a service or provision of products and profits depends finally on the assembly skills of those concerned. No design, specification, legal or financial aspect will be relevant without this and the ability of the main contractor, specialist engineering contractor and sub-contractor to work together. So perhaps the considerations now outlined are the most important.

Relative costs

Up to more than two thirds of most projects costs are for employees. Their organization, objectives and welfare are paramount factors in any project and will increase in importance as manpower costs and demands increase. Usually it is easier to overcome project technical and supply problems than those of human reactions. All contractors are familiar with the difficulties of a fluctuating workforce, and even when there is a trained management and supervision nucleus there will still be a need to employ local manpower and establish a method of working at each site and on each project. Labour relations will differ throughout the world and may be less important in some areas, but good morale, motivation and coordination will be essential, irrespective of UK manpower experiences.

Project planning

A contractor will need to establish a project plan. This will be as important for work control as the project design prepared by the consultant. It may be known as a network plan, a works table or a variety of other names, but its purposes irrespective of name are effective control and success. A suggested control structure is given in Fig. 5.1.

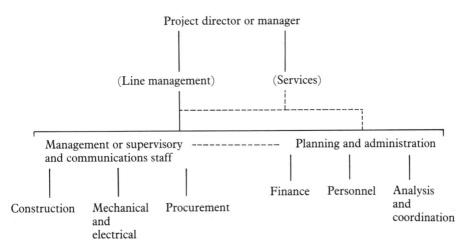

Figure 5.1   Project control structure

Objectives and communications

The purpose of any structure must be to define objectives, responsibilities and encourage good communications. For management of some projects a deputy project director or manager is suggested. This may seem extravagant, but under the stress of international conditions it will allow essential flexibility and forward planning which may otherwise be impossible. There are a number of precedents which are now accepted for this on construction sites and in factories and other organizations.

Organizational structure effectiveness will depend on the training and skills of those concerned, and above all on good communications. Dissemination of information and details of project needs and objectives will be essential. All should encourage and permit consultation. When possible this should go beyond the job site and involve suppliers and those responsible for freight, distribution and as many supporting functions as possible, particularly cash flow. It may be expensive and inconvenient to establish site training, but on large and lengthy projects this and other training may be essential, and may also be a condition imposed by those financing or approving the project.

Methods of working and characteristics of management

Much will depend on the existing management skills of the main contractors, and others, and the ability to impose these at each site. Difficulties arise in the UK, but they are far more evident on an international project. Organized manpower may have less influence than in the UK, but other problems will occur and will need early settlement without reference to a superior. Managers and supervisors must, therefore, often be more knowledgeable and self-reliant. They certainly must be decision-takers and have the diplomatic skills capable of resolving not only the problems of the main contractors, but also the conflicting problems of the specialist engineering contractors and sub-contractors. Throughout the profit motive is likely to predominate, but may not always be as enforceable as in the UK.

Another dominant factor is the scale of a project. Those at the lower end of the scale, say up to £4 m. are now very much within the province of local contractors. Above this, say to £10 m., external expertise may be necessary, particularly for projects involving advanced technology and when standards of finish or completion on time are critical. Advanced engineering, factories, desalination and petro-chemical plants, telecommunications, power generation and a number of public works installations can come in this category. Government finance may also be involved, and is certainly likely for major projects, say, exceeding £100 m. Provision of finance from an external source may also be a factor which will influence the conduct of the main contract and in turn the involvement of each contractor, supplier and consultant. Possibly the project will be a turnkey or package deal for which each has a particular contractual liability.

Government influence and sources of finance

Government-to-government relationships have grown in recent years for many trading activities, such as defence sales. Construction, too, has been influenced. Sales of expertise and equipment are aspects, but diplomatic prestige and long-term objectives are also important features. In turn these can influence the relationships of contractors. They can also be advantageous to contractors, and designers, and lead to government support and diplomatic activity at the pre-bid stage, during a contract and, often more importantly, during final phases when last payments are due. The measure of the importance of international construction to the UK can be seen in part by the attention which is given to it by the Foreign and Commonwealth Office and the Department of Trade, and by such support as grants towards costs of tendering. These costs are high for medium-size projects up to, say £50 m., and can be prohibitive for large-scale projects which may range from £50 m. to £1 000 m. or higher. Exports credits guarantees are another important support. All can be linked with financial assistance given to a foreign country.

The source of finance will be a key factor influencing how a main, specialist or sub-contractor assesses a price for a job and the risks involved, and it will certainly determine how work by a contractor can be financed and the costs. Payments are always behind costs incurred. International competition is increasingly putting more power in the hands of buyers and employers, and when public sector finance is involved, probably with diplomatic factors, the power exerted by the buyer can be excessive and trap less experienced contractors into low bids. To be successful this results in later pressures being put on suppliers and others to ensure even the lowest of profit margins. Acceptance of bids at unreasonably low prices can lead to bankruptcies, delayed payments and a general weakening of quality and abilities to perform. Regrettably, some international contractor is usually prepared to succumb to this.

Assessing financial risks and profit potential

Each contractor must assess the financial risk and eventual profit in relation to a bid or part of a project. Few contracts of any scale run for less than a year. Many run for two or more years and require the most careful study of financial implications. Even if the will, expertise and management are available for over-

Table 5.1 Assessing financial risks and profits of a project

Function	Time in months	Cost as a percentage of contract bid price
Concept and feasibility study	6–18	1
Negotiation with the consultant, suppliers, bankers and insurers, and pricing	12	0.5 to 1
Design work and approvals	6–18	5 to 7
Construction	18	20
Sub-contractors and suppliers	18	35.0
Freight	12	3.0
Commissioning	6	3
Project management	48	4 to 5
		75
Insurances including exports credits		1 to 2
Financing, taxes and other costs	—	10 to 15
		92
Profit		8
Total		100

seas projects, cash may not be. Costs involved, resources, inflation and other factors must govern the building-up of an assessment of financial risks and profits. An example of this is given in Table 5.1. Any assessment will be dominated by the need to meet performance at a stated price, usually with penalties for a failure to perform, and to obtain profits. It is emphasized that the figures given in the table are only examples. Much will depend on circumstances and how profits are sought at each stage of a project and by whom. A final profit of 8 per cent is clearly little for risks involved, but may be the best likely. Profits may, however, have been earned by associated companies at other stages which could make the final profit more acceptable. The cost of financing is the one factor likely to affect the project throughout and is one which the controlling staff is unlikely to be able to influence by economies, design changes or expertise.

Impact of contractual requirements

The function of each contractor will be influenced by contractual matters which will dominate every aspect of working irrespective of skill and experience. Acceptance of a contract or making a bid may be unacceptable if contractual conditions are unreasonable. Overseas work can be hazardous at the best of times. Unreasonable contractual conditions will do nothing to help performance and can lead to protracted legal arrangements or arbitration and consequent delays of payment, and not least to difficulties of relationships between contractors and others.

At each stage of planning the relationships between all interested parties will depend on the price quoted for a job, the expected profit margins, objectives and penalties. Discussion and disagreement may be expected between the commercial, technical and advisory or service personnel involved, but agreements are necessary if profits are to be made and if arguments over production failures or legal and financial problems are not to result. Decision-taking will be helped by preparation of detailed costings and the building-up of tables to assess financial risks. Such preparation, while common to many international contractors, may be less familiar to others. Even if becoming more familiar, much depends on the experience and abilities to anticipate and assess likely problems.

Those involved in a project may find it simpler to resolve technical and production problems than those which are based on commercial or legal factors. The demands of meeting targets and facts often dictate a production solution. Regrettably other matters are not so easily resolved, and result in attempts to transfer liabilities or penalties. Consultants, suppliers, main and sub-contractors and the employer may each be partly at fault, and contractual agreements are never likely to be rigid enough to provide accurately or fairly an answer. Unless there is goodwill between interested parties time-consuming and expensive arbitration could result. There can, therefore, be no substitute for the satisfactory selection of partners, suppliers or sub-contractors as a part of the working relationship or as part of a more structured partnership or joint venture. The objective must be always to ensure the type of relationship which provides for speedy, amicable and satisfactory decisions at the minimum cost. Competition is such that usually no interested party can afford to accept unreasonable liabilities above those already calculated.

Requirements for cooperation

Participation in, or even consideration of, a contract will be governed by its size. A main contractor may be supported by finance from within or from a government or similar source, and the very scale of liabilities will be a severe controlling factor which will prevent capricious bidding, apart from any bid bonds imposed. However, suppliers and sub-contractors may not be similarly restricted and the main contractors will face the demanding task of good selection. Price should not always be the controlling influence. Selection of suppliers will usually affect sub-contractors to a greater degree than others, and will depend on the specification or negotiation with the consultants. Good selection and performance of suppliers will dominate a project.

No single factor is likely to be more consistently disruptive than failures by suppliers, which will affect completion and penalties for all. Again much will depend on the experience of those involved. However, there are those who will bid for a contract with too little assessment of a specification, usually having had too little time for comparative study of requirements and frequently without knowledge of the manufacturing standards involved, but with the hope of overcoming any financial complications or shortfall by savings on another, often by pressures on suppliers. While common the results can be time consuming and lead to disruptions which can certainly reduce eventual profits.

Many problems result from the lack of time given by an employer or designer for completing a bid. It seems quite illogical that a complex design which may have been discussed in principle for a year or more, taken one, two or more years to design and is expected to take many months or years to complete under fluctuating economic, political and other conditions should be given to a contractor for a bid within at best only a few weeks. The result of this and similar unknowns is to lead to the necessity to build contingency costs into a financial plan. However, over-provision leads to the failure of a bid and under-provision to the failure of profits, or worse. Needless to say any figures quoted change with circumstances and objectives. Similarities between the costs of various parties may be expected, but these could differ significantly between countries. Some may face higher manpower costs than others, some manufacturing and supply problems. The financial basis of working may also be very varied. The essentials are, therefore, to ensure good analysis of a job in all its aspects, selection of management and manpower and the establishment of realistic objectives, and to ensure as far as possible that all can work together, or at the very least have the experience, will and intention to work together.

41

Chapter 6

Establishing a joint venture

Structured methods of working

Relationships between those involved in a project may be informal and based on methods of working established over many years. The relationships of some may be formalized contractually, but even so will only work effectively through mutual understanding and agreement. In many circumstances working relationships can be helped and encouraged by the establishment of a joint venture. The principles are as applicable to a manufacturing or selling activity as to a construction project. A joint venture is usually only viable when each of those involved has something to offer to the other or to share the risks of a contract. Each partner must contribute particular expertise. In some cases a local joint partner is essential to obtain a contract.

There is a variety of joint venture agreements, and, in most cases, the partners may have a joint and several liability to a customer, such as the joint venture where the partners pool resources and share resulting losses or profits. Alternatively the role of each partner may be defined and each take responsibility for a share of the work, but still be jointly and severally liable to the customer. There is also a type of joint venture in which a leading partner is the contracting partner to the customer and avoids other partners having a direct relationship with or liability to the customer while undertaking to perform their part in a joint venture agreement as being responsible to the leading partner and other partners. This method of working is particularly favoured by some European contractors.

The impetus to establish joint ventures has grown in recent years because costs have increased the investment required for projects beyond the resources of individual companies. International construction projects, particularly where there is a limited industrial infrastructure, have led to an increasing number of joint ventures being arranged to undertake projects. A further reason for joint ventures is the availability of credit. Joint ventures may be between main contractors, specialist engineering contractors, sub-contractors or between finance or development organizations and companies with particular expertise. They may involve companies from other countries with one of the partners being from the country in which the project is to be undertaken. Local partners may be required by law in some countries, which can cause difficulty.

Deciding on a joint venture

While joint ventures are usually associated with large-scale activities, they may be undertaken for all sizes of projects, but decisions to join with others must depend on the size, complexity and location of project and expertise available. Any firm may decide that it lacks particular experience, finance or management and can only undertake a project in partnership. There may be a need for participation with someone having local knowledge who is able to combine such expertise as estimating, knowledge of local procedures and resources which enables participation in projects otherwise excluded. A joint venture can extend capital resources or spread the risk, thereby giving greater bidding and bonding capacity and credit facilities. Shared equipment from the resources of more than one can encourage competition and reduce investment, but the emphasis must be on the combination of resources to give greater strength and competitiveness. Quite simply a joint venture should be considered as the pooling of assets and expertise of organizations to achieve an objective and share all responsibilities, liabilities and profits.

Methods of establishment

The way in which a joint venture is established depends on the needs of those concerned and local law. It may be established as a general or limited partnership or as a limited liability company. Apart from the partnership structures which can be used, there may be a need for different agreements when partners have similar expertise or to provide for a joint venture between companies and a professional partnership. It may be semi-permanent and should be established before tendering for a project.

A joint venture can differ from the accepted relationship between a main contractor and sub-contractor as it requires members to be jointly and severally bound to an employer. An alternative may be the formation of a sponsor company or a partnership arrangement, usually with a coordinating contractor being nominated. Variation of these can lead to a joint venture being created involving design consultants for the specific purpose of a combined bid for the design and construction of a project.

Key questions

A primary objective must be to provide the most effective relationship between all concerned, specifically to ensure that the parties have the expertise to meet commitments and, possibly, a history of working in a joint venture. At an early stage the title of the joint venture must be decided and the objective defined. For this details of the buyer must be obtained with a decision on the form of the agreement, possibly for pre-bid only, and its duration. The law of the main contract and that of the joint venture agreement may be different. This must determine a procedure for settlement of disputes under the main contract, should it be different under the joint venture agreement. A main contract may allow for differences under a joint venture agreement, but it may not. In fact the main contract may not provide for any matters to be included under a joint venture agreement. Factors for consideration when creating management for a joint venture team which can determine the type of joint venture, whether sponsorship company or partnership, or may follow decisions on the type to be used include: the proposed organization and needs of the parties; main contract restrictions on the form of organization; sponsorship responsibilities, obligations and liabilities; board of management or committee and frequency of meetings; employment of management staff and responsibility for specific matters; appointment and powers of the project manager; liability of the project manager to the parties of the joint venture and the appointment of other managers.

Sponsorship companies and partnerships

As the name implies, a sponsorship company is a legal entity formed to be responsible for the conduct of a project. This may be a usual choice when the work of a joint venture varies, when a number of companies are involved or when the decision for a joint venture has been taken to cover more than one project. If joint venture parties are from different countries a sponsor company provides a measure of efficient control of operations and separates management from the role of an individual partner. Usually such a company will be staffed by personnel seconded from the joint venture partners or by those recruited for the purpose. It should be provided with working capital by the partners and have its own management responsible as a company to the joint venture as partners. It is essential that the overhead costs of the company be kept to a minimum, and it is a matter of assessment to determine needs for operating capital in relation to a competitive tender.

A separate company may be created, and the joint venture partners may create a structure within which management decisions on behalf of the joint venture can be taken. This may be in the form of an executive committee, but one of the partners should be designated as the coordinating partner responsible for the coordination of work. This arrangement is more appropriate when the partners are undertaking the work with a limited amount of sub-contracting. It may still be possible to sub-contract, but in some cases there can be problems with the relationship of sub-contractors to the partnership which would not occur if the companies involved were sub-contractors to a sponsorship company.

Closed consortium

Another form of joint venture structure is that in which a leader is appointed, often a contractor, who accepts liabilities to an employer which are then passed on to others in the venture. This means that the contractor, sub-contractors, specialist contractors, suppliers or bankers involved can have an agreed share of liabilities or participation in profits which in turn can dictate a build-up of a financial contribution. Since the leader deals directly with the buyers in this structure it is possible to avoid joint and several liability. Clearly there are advantages in the structure, not least of which is in the simplification of providing finance by each partner rather than for the venture as a whole.

There will still be a need for coordination, particularly of the distribution of profits, but day-to-day finance will also require attention. For the greater part of the activities of the joint venture the roles specified for each partner and agreed financial participation can provide adequate control as profits will depend on targets being met. Such management by objectives protects the main contractor or leader, who can also obtain further protection by the establishment of a reserve or indemnity fund.

Allocation of responsibilities

There must be an early determination of the responsibilities of all parties in a joint venture. This may be dictated by the relevant project, but even so there may be need to provide some flexibility, particularly in relation to the interests and skills of those involved and provision for changes of participation. Permission to sub-contract must also be defined. Of great importance is the definition of members' responsibilities to each other and to a buyer, which may be several or joint and several. The law of the country of operation may stipulate this, but if not a decision is needed.

Clearly when competitors join in joint venture problems of exchanging information not readily applicable to the joint venture project may arise. This could affect technical and commercial planning and lead to delays in a programme of work which place considerable responsibilities on the project manager. When a project is the forerunner of another or a basis for experience useful elsewhere, there may be great caution about exchanging information, even with the most carefully selected partners of any nationality. It must be expected, however, that there will be a pre-determination to joint venture between partners of the same nationality, but local laws may dictate otherwise. When specifying responsibilities there must, therefore, be confidentiality and a definition of to what, and to whom, this should apply and restrictions on what may or may not be given to third parties.

Tendering

One of the first aspects of a joint venture to be affected by definitions of responsibilities could be how prices for a tender are to be prepared. Work may

45

have to be divided between parties following pre-qualification. Characteristics of those involved will be disclosed. They will be evident in reactions to tender prices preparation and matters peculiar to the contract. If changes are to be pursued responsibilities and objectives must be agreed and stated and the apportionment of tender costs agreed. The latter is very important, but more so when a tender fails to result in a contract.

Financial planning and liabilities

Before or during the preparation of a tender agreements will be needed on financial aspects of a joint venture. Ideally these should have been settled prior to the establishment of the joint venture structure. Particularly relevant to preparing a tender will be responsibilities for providing bonds and the costs involved, insurance and methods of accounting for payments received from an employer by the joint venture. Working capital will be needed to allow the joint venture to function, and for the project. This may be provided in equal proportions or in other agreed amounts. Such amounts could determine shares in liabilities and profits, and certainly in the acceptance of additional costs or need to raise more capital.

Financial management is likely to be the responsibility of the joint venture manager. There must, therefore, be clear instructions on invoicing between the partners, accounting for work done, and financial management generally. These and the joint venture agreement will determine liabilities to the buyer, suppliers and other third parties, limitations of liabilities, settling disputes and, indeed, the liabilities of the joint venture manager. Each member of the joint venture will have a method of accounting for an investment in the venture, but this must also have a separate financial identity which provides for the matters outlined. Other routine factors which must be considered are: the management of accounts; accountancy procedures; preparation of accounts and financial reports; audits; methods of payments and cash flow controls; currency controls and exchange fluctuation and limitations.

Manufacturing and other activities

Joint ventures are by no means restricted to construction projects. Indeed, in developing countries and elsewhere there are long histories of joint ventures ranging from the development and extraction of natural resources, processing, manufacturing, distribution and marketing and selling. Design and research and development are more recent joint venture activities which are growing as expertise and experience increase. Establishing a joint venture for any of such activities may be approached from different viewpoints from contractors, and generally are likely to be longer-term. As such, market assessment, feasibility studies and planning generally may be more extensive, and take longer. A possible sequence of events which can be considered in relation to activities already outlined and market planning generally is the analysis of the manufacturing, service, selling or project function; decision to establish a joint venture; define objective; market research and planning; selection of partners and joint venture contract; local organization; market, manufacturing, service, selling or project plan; expansion planning and policy formation and training.

Disagreements and expulsions

Few partnerships are likely to be without disagreements, and in construction activities, which are a series of frustrations, disagreements are common. Fortunately, and surprisingly, they are overcome more easily and quicker than seems to be the case in many other industries. However, a joint venture which brings together different structures, skills and personalities must provide for disagreements, resignations and expulsions.

The insolvency of a member may prevent further participation, as would failure to meet a call for additional capital. The latter may, however, lead only to a reduced share in profits, and liabilities. Failure to meet obligations of any type may require expulsion on terms provided in an agreement, but key questions must be how will the venture continue after an expulsion and whether or not replacement is acceptable or possible, and how existing contributions can be wound-up, returned, transferred or continued.

Law and goodwill

Much will depend on the law governing a joint venture. Of equal importance is, neverthless, the attitude of buyer to the obligations of a joint venture. It follows, therefore, that selecting a partner, agreeing expertise, contributions, legal framework and the many aspects of a joint venture depend essentially on goodwill and intentions to succeed rather than laws.

Chapter 7

Financing construction activities

Sources of finance

Finance for design, manufacturing or construction work may come from a variety of sources. The World Bank, supranational bodies such as the Commission of the European Economic Communities, national governments, local authorities, public corporations or any one of numerous private undertakings or individuals may initiate a project or provide funds. Whatever the source, there will be a need for each individual partnership or firm to have initial working capital and financial controls.

There is no shortage of international construction work needed, only a lack of capital, or a lack of ability to balance financial resources against construction needs. Even when such problems are overcome, and payment for work or products is apparently assured, much will depend on the evaluation of financial and other abilities of those seeking finance by those responsible for its provision. Methods used by all may be very similar, although state needs can be influenced by factors which are irrelevant or of less concern to a private undertaking.

Raising working capital

For most private undertakings, whether or not quoted on stock exchanges, the role of banks will be significant. Whatever the issued capital of a company or partnership the size of activities will be dictated by the amount of working capital needed, available or on call. Companies, particularly, may have a financial structure based on ordinary shares, debenture stock, convertible stock, preference stock, bank loans or any combination of these and other financing methods. The provision of capital, whether from individual, company, bank or other sources and any evaluation will usually involve a bank. The principles which can govern obtaining bank finance are equally applicable to an assessment by an individual or group of persons holding funds or responsible for utilizing capital. A market assessment or feasibility study should have assessed or be in the process of assessing the experience, structure and ability to support overseas work, but financial planning will also test such an assessment and raise other questions. An outline of basic financial information which governs any business function is set out below.

Balance sheet

Debts and advance payments; cash due; work in progress; depreciation; investments and reserves.

Cash flow projections

Actual cash flow and projections for at least one year with further projections for two years or more; collections due; payments due; overhead costs and equipment expenditure planned or likely.

Profit and loss statement

Sources of revenues; income from joint ventures and similar activities and extraordinary receipts or income.

Financial strategy

Plans for financing existing and future work.

Cash flow

Financial information of the type outlined should enable a banker or board to determine whether an undertaking is over-extended, profitability of existing work and current and future liquidity. Working overseas may be hazardous owing to erratic payments, which places considerable pressures on cash flow. It is such questions as these which must be related to other market factors and determine whether the management of an undertaking has fully appreciated the complexities of overseas work or investment. Other financial factors are currency exchange risks, inflation and, very importantly, past or existing exposure to overseas activities. Experience in one country may not justify expansion to another although it will certainly justify consideration.

Much will depend on whether or not entry in an overseas market or additional overseas market will be an over-extension of commitments. If the market in one country is reducing pursuit of others may be essential. It may also be advantageous for contractors to obtain cost-plus contracts to offset fixed price contracts, or for designers or manufacturers to diversify to take advantage of a market likely to develop in the short or medium term. Their time scale from concept to production is usually longer than for a contractor who has obtained a contract, although obtaining this may have been an equally lengthy process. For each, however, the basic questions are similar and cover the ability of those concerned to remain solvent, how mobilization and logistical problems will be overcome and the management skills available to operate internationally. The latter is very much a matter of personal judgement and almost impossible to quantify, particularly in a financial statement. Many international activities may be pursued successfully against adverse factors on the strength of good and experienced management.

One of the major financial questions governing overseas work must be about the source of payment, followed by questions about willingness and ability to pay. Political risks must, therefore, be high on the list of factors to be assessed. Despite very detailed study at home these can never be fully interpreted until visits to the country concerned are made. These visits may need to be frequent and lengthy after initial studies at home, and are essential for all decision-taking, particularly when the provision of a bid bond, performance bond or performance guarantee is involved. Financial matters will need study by contractors to support any pre-qualification for a contract and apply equally to supporting the claim of a designer or manufacturer to particular experience or status. They must always relate to wider market assessments and may also be used as a basis for questioning the probable outcome of political and financial risks likely to affect a project. The most important are as follows:

Pre-qualification information

Projects completed and role in projects; corporate structure and organization; names and résumés of key personnel; equipment and references.

Current operations

Description of markets; projects, types and scope of work; local facilities; progress; prospects for projects; project review covering size, amount completed and paid for, profits to date and estimate of profits on completion.

Future plans

Markets being pursued; type of work to be undertaken and personnel and capital needs.

Changing financial sources

Providing finance for commerce and industry throughout the world has changed in recent years due to the reduction of private capital which is available and the consequent reliance on government, local government or other public or quasi-government sources for working capital. In many cases high costs of research, such as in the aircraft industry, or high development costs, such as for oil or minerals, have led to more and more reliance on government loans. The UK is as much affected as other Western European countries, and even those countries such as the USA, which have sought to avoid too much government financing unless it is for a limited number of industries influencing national strategy or prestige, are faced with the same problem. Developing countries' dependence on government finance is usually most marked, and even those with massive income from resources have found that government control is required to pursue objectives which depend as much on scarce expertise as on home revenue or that from world funds. However, irrespective of the source of funds the political or economic structures of developing countries are such that firms from industries like the UK hesitate to pursue markets unless their trading activities are underwritten in some way. This has led to provision of export credits guarantees from government or other sources. Facilities may be sought for capital projects and by designers, manufacturers or contractors. The main types available in the UK are supplier credit and buyer credit.

Supplier credit

Under a supplier credit arrangement an exporter gives a customer credit and in return receives from the customer a bill of exchange or promissory note which can be sold to a UK bank. The bank transaction is covered by a guarantee from the Export Credits Guarantee Department (ECGD). This and services provided by it and similar institutions are an essential part of exporting. Supplier credit is generally used for smaller contracts, and the credit period is normally of a limited time from delivery of the goods or commissioning of the project. In addition credit is usually limited to 80–85 per cent of the value of the UK goods and services to be supplied under the contract, but a proportion of goods and services from other EEC countries may be included within this limit. The services of ECGD in the UK and those of comparable bodies elsewhere are:

UK – Export Credits Guarantee Department (ECGD)

Guarantees to banks making loans, insurance policies covering political and commercial export risks.

France – Compagnie Française d'Assurance pour le Commerce Extérieur (COFACE)

Guarantees to banks making loans, insurance policies covering political and commercial export risks, interest subsidies.

The Netherlands – Netherlandsche Credietverzekering Maatschappij, NV (NCM)

Guarantees to banks making loans, insurance policies covering political and commercial export risks.

Federal Republic of Germany – Hermes Kreditversicherungs AG (HERMES)

Guarantees to banks making loans, insurance policies covering political and commercial export risks.

51

USA – Export Import Bank of the United States of America (EXIM)

Loans, guarantees to banks making loans, insurance covering political and commercial export risks.

Buyer credit

A buyer credit arrangement enables an exporter to negotiate a contract with a customer on a cash basis. At the same time the customer negotiates a loan agreement with a UK bank to assist making payments due to the exporter. In this case the bank also obtains guarantees from ECGD. Credit facilities are available for contracts from £1 m., but are generally used for larger contracts, say for those exceeding £2 m. For contracts below £2 m the credit period is limited. For larger contracts availability depends on the size of the contract, and could be over five years from commissioning. Large contracts can have a ten year term. However, terms for each contract are decided by ECGD and vary according to the country, type of contract involved and other factors. Credit for developed countries, especially EEC countries, is normally limited regardless of the size of the contract. Terms in excess of eight years are usually available only for the least developed countries. In these cases the amount of credit is also limited to 80–85 per cent of the value of UK exports, including a proportion of other EEC exports, but limited credit may be available for less developed countries for selected local costs. These may be repayable over a shorter period than for the main credit.

When using a buyer credit arrangement it is envisaged that finance is provided for a single contract concluded at the same time as a loan agreement. However, there may be occasions when a buyer wishes to place several contracts for one project, but not initially. In this case it may be possible to arrange a line of credit with a UK bank on a buyer credit basis, which enables contracts to be placed over a period. The length of credit could vary in relation to the value of contracts placed in the same way as an ordinary buyer credit, and the amount of credit available could also be similar.

Export credits terms and conditions

Export finance from the UK for capital projects has normally been made available in sterling, which may also have been the currency of the supply contract. However, the ECGD may issue guarantees for buyer credits in US dollars or West German marks. When such a credit is used, the supply contract will be in the same currency, but it should still be possible for the supply contract to remain in sterling. Interest is fixed for the life of the credit at a rate chosen by the ECGD. In 1978 the ECGD published details for determining rates of interest in line with an International Agreement on Guidelines for Officially Supported Export Credit which involved dividing buying countries into three categories. For credits of up to five years there are different minimum interest rates for wealthy countries and a different minimum rate for the others.

In addition to interest charges there are also bank charges. The amount and structure of these depend on whether or not the credit is provided in sterling or a foreign currency and prevailing rates, but they may be expected to raise the effective cost of credit by about half a per cent annually for a seven year credit. In a supplier credit they are payable by the exporter who provides for them in the contract price. The exporter also pays ECGD a credit insurance premium which varies with the length of the credit and the country concerned. This premium is included in the contract price and is not disclosed to the customer. Payments of interest are made half-yearly from the date of signing the credit, and payments of the principal sum are made half-yearly, the first payable six months after delivery or commissioning.

Practical implications of a buyer credit

There is no shortage of orders for construction contracts, design or products for those able to provide finance in addition to the products and services involved, which has led to the provision of financial packages being integral to obtaining a contract. Buyer credit can be provided to ensure this.

A designer, supplier or contractor can provide credit to a buyer by extending payment terms. In most cases invoices will not be retained and will be discounted. The person or organization undertaking this indirectly contributes to financing the contract concerned. However, when a contract involves substantial sums of money with payment terms extended over several years discounting is unlikely to be satisfactory as, among other things, it will affect the suppliers' balance sheet and prevent control by the discounting person or bank. A buyer credit is, therefore, an alternative, which funds being arranged with a bank for a buyer by the supplier. The loan obtained enables the buyer to repay the supplier over a period, which is clearly advantageous, and is a direct contract between the customer and a bank. With the growing strength of buyers this is an acceptable means of encouraging export sales, and is supported by the UK and other governments through export credits guarantees as already explained, thereby matching the needs of buyers who may be a risk with national financial support. The Commission of the EEC, among others, wishes to prevent such government support disrupting free competition as do most major trading countries. This resulted in the guidelines already mentioned, which are subscribed to by ECGD, COFACE, NCM, HERMES and EXIM among others.

The International Arrangement on Guidelines for Officially Supported Export Credit was ratified in 1978 by twenty-two OECD nations, and was designed to provide a framework for the financing of capital projects where long-term credit is necessary. The agreed level of interest rates for the credit has become increasingly unrealistic as market rates for borrowing have moved higher, and the distinction between aid and credits has become confused.

Some buyers, notably those with political pressures, often press for maximum buyer credit including local costs exceeding 80–85 per cent. Suppliers from countries without government support facilities may also be sought. When these factors are added to a low credit rating considerable financial expertise is needed to provide finance through bond issues, loans and government-supported loans. These may also involve the World Bank and other supranational financing bodies, but the essential factor is the provision of national or international guarantees to encourage private finance. The following are the types of contributions involved.

Buyer

- Payments or equity participation. This may be the difference between money raised and amount needed, or the difference between financing available or needed.

Bank

- Long-term loan with regular repayments and variable or fixed interest.
- Short-term loan with lump sum repayments and variable or fixed interest.
- Bond issue, lump sum repayments with fixed interest.

Government

- Loan.
- Guarantees loan made by bank.
- Insures against bad debts.

Contractor or supplier

- Guarantees credit.
- Guarantees difference between amount raised and amount needed.
- Interest subsidy to banks.
- Direct loan.

When preparing a manufacturing proposal or bid for a construction contract part of a submission may be a financial proposal. This will follow discussions with a bank or consortium and could lead to supplier, buyer and banker consultations. It is essential that if a sale is pursued on the basis of providing financing there must be a competitive proposal based on a clear understanding of the financial implications. Examples are as follows.

Contractor or supplier

- Seeks bank and government export agencies support for financing terms for a specific project.

- Provides bankers and buyer with information about costs and schedules.
- Undertakes project and has progress payment invoices certified and receives payment.

Banker

- Indicates amounts and terms of credit available for buyer.
- Arranges loans and credits for buyer.
- Disburses cash against certified progress payment invoices.
- Collects interest and principle sum from buyer.

Buyer

- Seeks indications of finance available.
- Negotiates terms of credit.
- Arranges for check on construction and certifies progress invoices.
- Repays money to banks and agencies.

Liquidity and financial controls

Financing any undertaking means essentially cash, cash flow or cash sources related to good financial management and controls. There can be no substitute for local financial information, and any partner must be able to provide this as much as sales, contacts and other guidance. Taxation is a feature, but so too are legal and insurance needs. A good financial reputation may overcome problems more readily than many other skills, but this can only be sustained by effective accounting, accounting methods and discipline as without these those responsible from the site or factory to the board will either lose control or be unable to maintain effective supervision. Losses or lack of profits are the likely results.

Experience or an ability to assess local currency, pricing and financial features already explained is essential. This means having the knowledge to pursue government and other loans available and early, frequent, informed and open discussions with bankers and other financial advisers and those able or willing to provide finance. Above all it is necessary to calculate the cost of borrowing and to relate this to current costs, inflation and profit targets. Always calculate the costs of remaining liquid and the costs of buying work. This may be justified as a short-term or promotional measure, but in international construction with the high degree of involvement of governments, national prestige and need of many to earn foreign currency or maintain home sales, there is always likely to be someone willing to reduce profit margins to a minimum. For some this can mean failure. Avoiding the pitfalls involved depends on expert financial assessment and management.

Chapter 8

Guarantees and performance bonds

A price on credibility and performance

Needs to balance commercial and technical judgement with legal and financial demands are likely to be most emphasized when assessing the requirements of a potential buyer for guarantees and bonds, which usually lead to extra costs. With such a variety of skills and organizations from many countries involved in world construction markets it is not surprising that there is such a need, also because many customers may be new to industrial responsibilities and to construction purchases and management in particular and see guarantees and bonds as a necessary protection. Some construction organizations may have established relationships in countries over a period which has led to mutual understanding and confidence which has obviated the need for guarantees and bonds. Even so sudden economic or political changes could disrupt such relationships and enforce their need. There are some who resist providing the indemnities sought and insist that there are other ways to meet the needs of a buyer. There are also those who accept the need to provide the guarantees and bonds without a resulting commitment to insurance, the costs of which increase a bid price. This can be hazardous for the inexperienced, and it is usually only a likely course when those involved have considerable financial strength and experience with the country and type of project concerned. For most involved in international construction there will be constant need to understand requirements for guarantees and bonds, the principles and problems involved and their impact on bids and the work which follows.

Types of guarantees and bonds

There are a number of guarantees and bonds which are demanded. Each has a different framework and is intended to meet a particular set of conditions. They have, nevertheless, factors in common. The main types are bid bonds, performance guarantees, advance payment bonds and retention bonds. They are so important to world construction that the International Chamber of Commerce (ICC) through its Commission on International Commercial Practice and the Commission on Banking Technique and Practice has laid down uniform rules which should be observed. Known as Uniform Rules and Contract Guarantees, the objective is to make it possible for international trade to develop and ensure an equitable balance of the interests of the various parties concerned. They cannot prevent the use of on-demand bonds, but, as the size of guarantees demanded is high, the rules seek to ensure that a claim is only honoured when a beneficiary has a legal right to make a claim based on failure by a principal to perform obligations. This can be achieved without interfering with the rights of buyers by introducing conditions into a guarantee providing for the submission of evidence of any breach giving rights and of the value of such rights. Rules providing for this are thought not to be contrary to the purpose for which a guarantee is given, and a contract and the guarantee should state these conditions. They should also establish minimum acceptable conditions and not attempt to deal with recourse by a guarantor against a principal. It is clear, however, that these objectives will depend on discussions between the parties concerned. The definitions of bonds and guarantees are set out below.

Bid bonds

Cover the period from the submission of a tender until its acceptance or rejection.

Performance guarantees or bonds

Cover the period of supply or work on a project.

Advance payment bonds

Cover payments made in advance by a buyer to meet mobilization costs of a supplier or contractor.

Retention bonds

These are used in place of the retention of a sum of money to cover the period until the end of the maintenance period stipulated in a supply or construction project.

Bid bonds

Those initiating a tender are rightly concerned to ensure that bids received are serious, well prepared and that any resulting contract would be capable of execution. Regrettably some potential customers invite too many bids and may not receive the best services as a result. The objective of a bid bond, sometimes referred to as a tender bond, is to prevent loss if a successful tenderer fails to proceed, such as by not signing a contract, or by failing to give a performance bond or other guarantee. In the case of contracting this type of bond arose from the open tendering system when any contractor was able to submit a tender for a proposed project. This made it difficult to assess all tenders and abilities to undertake a project. In many cases a bid bond is an unnecessary protection. If pre-qualifying procedures are carried out properly, particularly by experienced and reputable firms, they should not be needed. Among other things open tendering can delay the award of a contract, and unrestricted invitations lead to lengthy and costly estimating.

Selective tendering is a better procedure, but depends on information for pre-qualification. The process enables those concerned to establish which contractors have the financial, technical and other resources to undertake a project and are suitable to be pre-qualified as potential tenderers. Once a list has been prepared competition results between those who have pre-qualified and are invited to tender, but the risk of unsuitable contractors submitting a tender is largely removed.

Provision of a bond always involves expense, and once given should be released as soon as possible. Any demand for a bond must be reasonable. If too high it could prevent bids from some. The bond must also state clearly the conditions under which it is applicable and the period of operation. This could be significant if a buyer wishes to extend the period of consideration of a tender or time for preparation as lengthy delays could preclude pursuit of work elsewhere. As in so many aspects of international work, much will depend on consultation.

Performance guarantees or bonds

Provision of a bid bond may not cause too much financial limitation on an organization, althought it must have serious study in view of its implications. However, provision of a performance guarantee bond is likely to have considerable impact on cash liquidity, bank facilities, pricing, profits and, indeed, every aspect of undertaking a project to its final profit. A bond may be provided by a surety company or insurance company and, like a bank guarantee, is a guarantee by one party to another for the performance of a third party.

On-demand bank guarantees are an unconditional promise by a bank to pay a sum of money to a buyer on request. The bank must pay the sum involved in the guarantee irrespective of the merits of the claim and whether or not there has been any breach of the tender conditions or contract conditions. Such guarantees are unpopular with contractors and others, but many bankers prefer this type of guarantee as it is unconditional and removes a bank from disputes. The danger of an unjustified claim is evident, particularly when there are political factors involved.

There is a difference between a contract for supplying products and for a construction project. A construction contract is for the supply of goods and services carried out over a period of time and must state in detail all contractual relationships. Such detail is unlikely when supplying goods and services.

Conditions applied in a performance bond refer to the relevant construction contract, and providing the conditions of contract have been prepared properly a contractor is protected unless guilty of default. In this case commercial conditions would lead to a penalty. Acceptable conditions under which such a bond is called should be reference to arbitration or an agreement on the amount involved due to the default.

There are differences in practice in some countries. As an example there is a market for surety bonds in the USA. Surety enables a buyer to look to a surety company to complete supply or a project on the same terms as stipulated in a contract prior to any default. This may lead to another contractor completing the work. It is not possible for a surety bond to be called unless there has been a proven default. A disadvantage of such bonds is the limited market.

Performance bonds may be written for a lower figure than for surety bonds, often at 10 per cent of the contract amount. This is a similar level to the amount which may be provided by a bank guarantee. As in the case of a surety bond the organization issuing a performance bond may take the responsibility for a contract following default by a contractor or supplier. The mutual interests are, therefore, evident and are much closer than in the case of an unconditional bank guarantee. Provision of a performance bond, whether on-demand or conditional depends on the commercial and technical approach to a project and, following from this, on financial judgement and proper legal advice. Important factors which must be considered are the ability to meet the demands of a project, payment of damages on failure to do so and the period of liability. While performance bonds are more usually associated with and sought from construction contractors they are, nevertheless, an important part of selling expertise and products of all types to ensure delivery and prevent other aspects of construction or manufacturing from being disrupted. Those with a procurement contract can be particularly affected. The provision of a bond and its drafting must therefore be as much considered by manufacturers as others in the construction industry. A basis for a bond is illustrated on p. 60.

Advance payment bonds

Suppliers or contractors may be given an advance payment to assist manufacture or mobilization costs for a new contract and particularly for the purchase of plant and equipment. Such advance payments may be secured by a bond to ensure that the sums advanced are used properly. The amount of the bond should reduce at stages of completion of the project and to a time scale set when the bond is issued. As payments are usually spent in the early stages of a project, when mobilization costs are being met, these should also be reduced or eliminated as expenditure occurs. There is, therefore, less resistance to an on-demand advance payment bond than to on-demand performance bonds. An outline of the wording for a bank guarantee protecting an advance payment is given on p. 61.

Form of bond

By this bond We ....
of ....
whose registered office address is at ....
.... (hereinafter called 'the ....'
and ....
.... whose registered office address is at
.... (hereinafter called 'the Surety/Sureties')
are held and firmly bound unto ....
.... (hereinafter called 'the ....
in the sum of ....
(.......................................) for the payment of which sum the ....
and the Surety/Sureties bind themselves their successors and assigns jointly and
severally by these presents.

Sealed with our respective seals and dated this .... day
of ....

Whereas the .... by an Agreement made between the .... of
the one part and the .... of the other part has entered into a Contract
(hereinafter called 'the said Contract') for the ....
....

Now the condition of the above-written Bond is such that if the ....
shall duly perform and observe all the terms provisions conditions and stipulations of the said Contract on the .... 's part to be performed and
observed according to the true purport intent and meaning thereof or if on default by the .... the Surety/Sureties shall satisfy and discharge the damages sustained by the .... thereby up to the amount of the above-written
Bond then this obligation shall be null and void but otherwise shall remain in
full force and effect but no alteration in terms of the said Contract made by
agreement between the .... and the .... or in the extent or
nature of ....
.... under the said Contract nor any
forebearance or forgiveness in or in respect of any matter or thing concerning
the said Contract on the part of the .... shall in any way release the
Surety/Sureties from any liability under the above-written Bond.

Signed Sealed and Delivered by the said ....
in the presence of:

The Common Seal of
was hereunder affixed in the presence of:

Signed Sealed and Delivered by the said Surety/
Sureties in the presence of:

The Common Seal of
was hereunto affixed in the presence of:

Advance payment bank guarantee

WE refer to .... signed on the .... day of ....
Between:

Whereas
The .... has agreed to pay to .... according to .... of
the said .... an advance payment of .... (...............................
....)
representing   of the   Price.

Now we the undersigned ....
whose Head Office is at ....
do hereby *undertake* to guarantee and pay any monies up to the amount of
.... (...)
on the first written demand of the .... by "registered" post duly addressed to the Head Office of the Surety and without prejudice to his rights hereunder and without it being a condition precedent to his being entitled to the payment under the terms hereof. Such written demand shall specify the general
grounds for his holding that the .... has failed in the performance of the
said ....

The Present Guarantee will come into force at the date of the receipt by the
.... of the amount
of the said ....

The .... will give an immediate advice to ....
....of the said payment.

The amount of our Guarantee will automatically be reduced according to
.... of the .... proportionally to the monthly payment certificates it being understood that it will be entirely amortized at the issuance of
the Certificate of Completion.

It will be released at the issuance of the said Certificate of Completion and not
later than .... whereupon it will automatically expire and become null
and void without it being necessary to return the Guarantee for cancellation.

In witness whereof the ....
hereby executes these presents his .... day of ....

Signed by ....
In the presence of ....
                    (   )

Signed by ....
in the presence of ....
                    (   )

Retention bonds

Purchasers of capital equipment, those employing contractors and some who buy products may be expected to insist on some form of protection for a period immediately following supply or completion of a project. This may also govern the period when maintenance must be provided by the contractor. A retention fund or bond could be used. In the case of a fund the amounts retained vary, but are usually between 5 and 10 per cent. The purpose is to meet any difficulties that arise which require further work and, therefore, expense. Capital is committed and a retention bond to assure that there will be resources for any work which is undertaken is often preferred.

A necessary expense

Most buyers will require that their purchase be protected. Guarantees or bonds are, therefore, inevitable, so too is insuring against them being called, the costs involved and the consequent impact on a financial structure. All costs must be absorbed in the price of a project, and the degree to which cover is provided or sought is critical in a highly competitive market. Government support can be helpful or, indeed, essential for large projects, and it is often necessary to consider the type of support available to or likely to be given to a foreign competitor as this could distort a free market bid and influence decision-taking in pursuit of an order or project.

Chapter 9

Export Credits Guarantee Department. UK Government insurance facilities

Protection for risk-taking

Industrial countries have always relied on export markets, and increasingly so in the past twenty years. Higher energy, minerals and commodities prices which are to some extent linked to the independence and growing status and influence of countries, which were captive markets until quite recently, have put pressures on the payments balances of most countries, thereby accentuating the need for exports.

Unfortunately many countries with mineral and other resources have not developed their sources of revenue and may, in addition, be poor risks owing to inadequate economic and political experience or other factors. Their needs for construction may, therefore, be unattractive markets for those in the UK and elsewhere, despite wishes to enter new markets, without some form of protection against loss. This is not always readily available from private sources even when political factors encourage a government such as that of the UK to give support. Provision of exports credits guarantees through a national body can help provide government financial assistance to a third party or government, possibly for diplomatic or trade reasons, help a developing country or one facing particular problems and provide a market for UK firms thereby encouraging employment. Most industrial countries have such a body.

Export Credits Guarantee Department

In the UK insuring exports and overseas investment risks can be undertaken by the Export Credits Guarantee Department (ECGD), apart from insuring in the private market.

The ECGD is a department of the government. It assists exporters of goods and services by insuring against the risk of not being paid and provides unconditional guarantees of 100 per cent repayable to banks. On this security banks provide finance to exporters. The EGCD also insures new investment overseas against the risks of war, expropriation and the restriction of remittances, and provides protection against part of the increases in UK costs for large capital goods contracts with long manufacturing periods. For major contracts it supports the issue of performance bonds, and for members of a UK consortium

63

provides protection from loss due to the insolvency of a member of the consortium. Although a government department the credit insurance operations operate as a business, balancing income and outgoings and applying normal insurance principles.

Export trade is grouped as trade of a repetitive type, involving standard or near standard goods, for which cover is provided on a comprehensive basis, and for projects and large capital goods business of a non-repetitive nature, which needs a specific policy for each contract. Cover for specific insurance is given according to how the credit is provided. In the case of supplier credit the manufacturer sells on deferred payment terms and borrows from a UK bank to finance the period from shipment of the goods until payment is received. The ECGD Department insures the exporter and may give a guarantee direct to the bank. In the case of buyer credit the exporter receives payment from the buyer, who draws on a loan from a UK bank to provide the payment, the loan being repaid by instalments. The ECDG Department guarantees the bank repayment by the overseas customer. There are in addition special types of cover to meet particular needs. Apart from insuring against the risks of non-payment and encouraging new markets without fear of a severe loss, ECGD support for export finance helps an exporter to offer competitive terms and win contracts that might otherwise be lost. Comprehensive insurance is applied to the insurance of consumer and engineering goods and to services when business is continuous, but within this framework individual needs can be met. This principle cannot be applied to contracts for the supply of capital goods or for construction projects as such contracts may not be followed by others in that market. These contracts are, therefore, underwritten individually, insurance being arranged at the time the contract is negotiated. The cover is similar to that provided by comprehensive policies. As specific cover does not oblige the exporter to offer any other business for insurance, who may select only this worst business, this leads to higher premium rates to keep such business self-supporting.

The ECGD does not provide finance for export credit directly, but facilitates lending by the banks, and has developed a range of direct guarantees to banks providing export finance. By arrangement with the government banks lend against the security of these guarantees at special interest rates, at half a per cent over base rate for credit of up to two years from shipment and at special fixed rates for longer terms. For a major project and capital goods business with contract values of over £1 m., ECGD supports loans made by UK banks direct to an overseas borrower. Finance is provided at a fixed preferential interest rate for up to 85 per cent of the value of the contract, and is repayable on terms related to the size of a contract. The exporter is also able to negotiate a cash contract and progress payments prior to delivery. The UK bank makes payments on behalf of the buyer from the loan, in accordance with the terms of the contract. The buyer then repays the loan by instalments, as stipulated in a loan agreement with the bank.

Lines of credit may be guaranteed to overseas borrowers to facilitate the placing of orders for UK capital goods. These are available to all exporters of plant and equipment, and ensure that the exporter is paid on cash terms. The minimum contract value may be as low as £10 000. The loan is usually for 80–85 per cent of the value of each contract and is repayable over two to five years according to the value of each contract.

Underwriting principles

To be effective and balance income and expenditure ECGD must use sound underwriting principles. In the case of contracts for the sale of capital goods or major projects, the financial position and business reputation of a buyer are investigated. Contracts for the sale of engineering products on terms of more than six months are treated similarly. Reports from government sources, such as embassies and high commissions, may help assess economic and political risks in a particular area, but all commercial intelligence is important for analysing risks. As in the private sector market a policy-holder is required to carry a part of any loss. The percentage which is issued by ECGD has been increased to 90 per cent for buyer risks and 90–95 per cent for market risks, leaving a very small risk for the insurer, who must, nevertheless, protect the interests of the business concerned and of ECGD. When fixing premium rates ECGD tries to avoid loss and surpluses on its trading account, but does not subsidize one exporter at the expense of another. Premiums for short-term business are paid annually and monthly premiums on declaration of business. Rates for specific business and for comprehensive business covered under the Supplemental Extended Terms Guarantee are determined for each contract based on the length of time of the risks and market in which the business is done. No annual premium is charged.

When a claim is paid an exporter is relieved of most of a loss, but there is still responsibility for recovery action which the exporter is best placed to pursue. However, ECGD may offer help and advice, particularly when the employment of a lawyer or collector agent is necessary. An exporter is, nevertheless, responsible for meeting the costs of debt collection before or after the payment of a claim, but once liability for loss is established ECGD may participate in costs. If a buyer counterclaims ECGD expects the exporter to resolve the matter.

The ECGD ensures that transactions are confidential, and requires that the policyholder does not disclose insurance cover as it is undesirable that a buyer should know that the supplier will not suffer from a claim. Although for buyer credit the buyer or borrower must know of the ECGD involvement.

Buyers seek the best terms available and there is, therefore, a danger of excessive competition between sellers in different countries to secure business. This can be damaging to the balance of payments of the countries concerned. As a result there are international organizations which consider credit terms to prevent such competition, such as the Berne Union, the Organisation for Economic Co-operation and Development (OECD) and the EEC. This has resulted in the adoption of guidelines which set minimum payments, interest rates and maximum credit periods for officially supported export credits on terms of two years or more. For using the guidelines buying countries are in three categories on the basis of income. Harmonization of exports credits is being considered by the EEC, but decisions have yet to be reached. There are, nevertheless, reciprocal arrangements within the EEC by which insurance given to a main contractor for supplier or buyer credit transactions for capital and engineering goods can include sub-contracts placed in another EEC country for up to 40 per cent of the value of the main contract. Similar arrangements have been made for inclusion of up to 30 per cent value with Austria, Norway, Sweden and Switzerland.

The ECGD policy is to put UK exporters on equal terms with any foreign competitor while avoiding unnecessary credit terms, but applying this creates problems. Usually credit of six months is given for raw materials and semi-manufactured and consumer goods. Credit for other products does not follow a precise pattern, but ECGD agree terms exceeding five years credit depending on the size of the order and the goods concerned. Only for very large projects will ECGD agree terms exceeding five years, unless such terms are also offered by a foreign competitor. To help exporters of capital goods and contractors for major overseas projects ECGD may offer cover for unavoidable increases in UK costs and guarantees to encourage the issue of performance bonds. The former is restricted to contracts of £2 m. upwards and to the second to contracts of £1 m. or above. Guarantee, for pre-shipment finance may be made available for selected contracts valued at over £1 m.

Insurance. Supplier credit

The Comprehensive Short term Guarantee

The Comprehensive Short Term Guarantee provides a comprehensive insurance for a volume of sales on credit terms up to six months at a low premium. Usually the policyholder undertakes to insure a total export turnover for a period of not less than twelve months. The ECGD may accept business on a range of markets selected by the policyholder if business offered provides a reasonable spread of risk, but higher premiums are charged when a policyholder selects markets. It is a continuous guarantee, subject to annual renewal, available to manufacturers, merchants and confirmers. Insurance relates to cover for non-payment to a policyholder in respect of goods dispatched overseas under a contract of sale made with a buyer outside the UK, and also makes provision for cover from the date the contract of sale is made. Cover is limited to goods to be exported within twelve months of the date of contract. The ECGD is not informed of individual transactions. There are, however, credit limits to the business for individual buyers which ECGD will insure. The type of risks and covers are:

- Insolvency of the buyer   90%
- Failure to pay within six months of due date for
  goods accepted   90%

66

- Failure to take up goods which
  have been dispatched
- A general moratorium on external debt decreed
  by the government of the buyer's country or of
  a third country through which payment must
  be made
- Action by the government of the buyer's
  country which prevents performance of the
  contract in whole or in part
- Political events, economic difficulties,
  legislative or administrative measures arising
  outside the UK which prevent or delay the
  transfer of payments or deposits made in
  respect of the contract
- Legal discharge of a debt, but not legal
  discharge under the proper law of the contract,
  in a foreign currency, which results in a
  shortfall at the date of transfer
- War and certain other events preventing
  performance of the contract provided that the
  event is not one normally insured with
  commercial insurers
- Cancellation or non-renewal of a UK export
  licence or the prohibition or restriction on export
  of goods from the UK by law

20% to exporter.
90% of balance, ECGD.

95%.

95%.

95%.

95%.

95%.

95%.

The standard guarantee covers all types of contract which relate to the export of goods from the UK. It can also cover re-export of goods imported into the UK, but there is a short-list of imported goods directly competitive with UK trade which are not covered. Cover applies to contracts made in other approved currencies listed in a schedule to the guarantee. Contracts made in other currencies may be covered by agreement with ECGD. When an exporter invoices in certain foreign currencies and protects against adverse fluctuations of the exchange rate by use of the forward exchange market or foreign borrowing ECGD will pay up to 10 per cent more, provided the exporter has suffered the extra loss through participation in the forward exchange market or foreign borrowing and is a holder of a Foreign Currency Contracts Endorsement.

The Supplemental Extended Terms Guarantee

Export of commercial vehicles, machine tools, machinery, contractor's plant and many engineering goods may need the grant of credit to a buyer in excess of six months. This may be granted by the supplier, through a UK merchant, confirming house or finance house. The Supplemental Extended Terms Guarantee provides insurances for such business. It is also a form of insurance for covering any continuous business when pre-credit risk cover is required and when the delivery period exceeds twelve months. It is available only to those who already hold a Comprehenseve Short Term Guarantee. The policyholder must offer all business for cover, or business in selected markets. If the latter selection must be the same as under the Comprehensive Short Term Guarantee. The policyholder is not permitted to insure extended terms business in markets which are excluded from the Comprehensive Short Term Guarantee. The Supplemental Extended Terms Guarantee is intended to cover individual contracts made with buyers overseas and continuous business with overseas distributors and dealers.

Each contract or cover must be approved by ECGD in writing. Cover is normally limited to contracts where the delivery period does not exceed two years and the credit period five years. A premium is assessed in a different way from the Comprehensive Short Term Guarantee. There is no annual premium or separate pre-credit risk premium. Each is determined according to the period for which ECGD is at risk and the grading given to a buyer's country. In other respects the risks covered and the form which the cover takes are similar to the Comprehensive Short Term Guarantee. The standard cover makes provisions for sales to subsidiary and associated companies overseas, or sales from overseas stocks when required. There is also a pre-credit risk section to provide cover from date of contract instead of from date of dispatch. Protection against foreign currency fluctuations is also available.

Subsidiaries Guarantee

The ECGD covers the sale by an overseas subsidiary of goods sold to it by its parent or associated companies in the UK, on short credit terms or extended terms when these are appropriate, but normally insists on covering the sale to the overseas subsidiary before covering the sale by the subsidiary. The Subsidiaries Guarantee is a separate guarantee which follows the general form and content of the Comprehensive Short Term Guarantee. A supplemental extended terms facility is also available.

The Supplementary Stocks Guarantee

Apart from covering sales from stock held overseas under the standard Comprehensive Short Term Guarantee, ECGD will also cover goods while held in stock against loss arising from such events as war between the UK and the country where the stocks of goods are held, requisitioning or confiscation of the goods by the government of that country or the government of a country through which the goods are in transit to stock and prevention of the export of goods from a holding country. The cover is given under a Supplementary Stocks Guarantee and is available only to holders of a Comprehensive Short Term Guarantee who have cover for the sales from the stock.

External Trade Guarantee

When dealing in foreign goods cover is available to UK manufacturers, merchants and others for transactions when goods are shipped from a supplying country to a buying country, but are not first imported into the UK. There are exceptions for goods directly competitive with UK trade, for which this cover is not available. The cover is by an External Trade Guarantee which is comprehensive and generally follows the Comprehensive Short Term Guarantee. The policyholder must offer all external trade turnover for cover or an acceptable proportion of the business. The risks covered are also similar, but cover is not given for the failure of a buyer to take up goods nor against the imposition of export or import licensing or the cancellation or non-renewal of licences previously issued.

Specific Guarantees

Transactions involving capital goods or projects on a scale that cannot be covered by continuing insurance may be covered after individual negotiation of insurance by the exporter and ECGD to obtain cover under a Specific Guarantee. This may be available from either the date of contract or the date of shipment. The risks covered are similar to those covered under comprehensive policies, but the top percentage of cover is 90 per cent. No cover is given against the failure of private buyers to take up exported goods. For sales of capital goods made to government buyers ECGD will cover the risk of default by the buyer at any stage in the transaction. Erection costs can be included when not covered by a separate policy.

Cover may be for credit up to five years under specific policies, and longer terms may be agreed. Specific policies are negotiated at the same time as the contracts to which they apply. At an early stage in the negotiations ECGD may confirm that cover is available, check the credit-worthiness, stipulate any guarantees of payment which may be required, state the proposed credit terms and indicate the likely premium. When negotiations enter a final stage ECGD will provide a firm offer of cover, normally valid for three months. Acceptance within that period binds ECGD to issue a policy in the specified terms. When cover is accepted, the premium on the whole contract is payable. Cover for foreign currency fluctuations may also be included.

Constructional Works Guarantee

The value of overseas construction earnings by UK firms is no less important than those in other industries. A construction contract provides for the supply of goods and the performance of services, and in this way is different from others. Both are covered under the ECGD Constructional Works Guarantee. Cover relates to any sums due under a contract. The main policy relates to contracts with government bodies, and provides 90 per cent cover on losses. Cover for contracts with private employers is also provided, but in place of the cover on default of the government employer there is cover on 90 per cent of a loss in respect of insolvency or protracted default on sums due under the terms of the contract, and on 90 per cent of loss from delay in sterling transfer of such payments. Dates for settlement of claims follow the same pattern as those under comprehensive policies. Risks covered are shown below.

- Default of the government employer (including delay of the transfer of sterling payments).
- War between the employer's country and the UK.
- War, revolution or similar disturbance in the employer's country.
- Imposition of import or export licensing or cancellation of an existing licence, for goods or materials manufactured or purchased by the contractor after the date of a contract, for use on the contract, but for which at date of loss the employer has no obligation to pay under the contract.
- Additional handling, transport, or insurance charges due to interruption or diversion of voyage on goods or materials shipped from the UK if these charges cannot be recovered from the employer.
- Within limits agreed with the contractor the employer's failure to pay to the contractor sums awarded in arbitration proceedings under the contract.

Cover is given on an amount which includes price increases, provisional sums and interest, which may also provide for the extra-contractual element in any arbitration award. The premium, payable at the outset, is charged on the total of the estimated basic contract price and all such additional sums, with provision for a proportional refund of a premium when the actual contract price and interest charges fall short of the estimate. An example of how the Constructional Works Guarantee relates to those involved in a typical project is shown below.

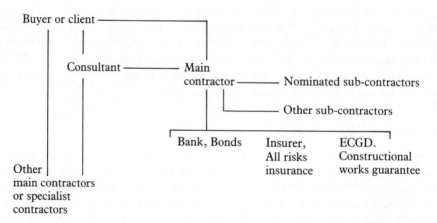

Figure 9.1 Constructional works guarantee. Relationships in a contract

Services policies

Cover may be needed by some sectors of the construction industry, notably consultants, for earnings from services for overseas clients in the form of technical or professional assistance, processing or hiring, or supply of expertise. Provided that the services are performed overseas or the benefit of services performed in the UK is gained overseas by the client cover is available under the Comprehensive Services Rendered Guarantee which provides the equivalent of basic comprehensive cover. Alternatively, a type of basic comprehensive cover is available under Comprehensive Services Sums Due Guarantee. Risks covered by this and the Comprehensive Services Rendered Guarantee, which excludes additional handling, transport or insurance charges on materials needed, are:

- Insolvency of the client.
- Failure to pay within six months of due date any sum payable under the terms of the contract.
- Government action which blocks or delays the transfer of sterling.
- War between the client's country and the UK.
- War, revolution or similar disturbance in the client's country.
- Additional handling, transport, or insurance charges on materials needed for the service and exported from the UK, arising from interruption or diversion of voyage if these charges cannot be recovered from the client.
- Any other cause preventing payment by the client which is beyond the control of the policyholder and the client, and occurs outside the UK.
- Repudiation of contract in cases where ECGD agrees that the client has government status.

Guarantees for supplier credit financing

Overseas trade may be encouraged by giving credit as part of an overdraft facility or by a bank advance against the notes or acceptances of a buyer. The latter may also be discounted. To assist, ECGD will agree to an exporter giving rights under a policy to a bank, either in respect of a whole policy, for all transactions in specified markets, or for all transactions with a named buyer. In some cases policyholders prefer to obtain a direct guarantee for their bank. This type of guarantee is an extremely helpful service.

Comprehensive Bill Guarantees

When a credit period is less than two years from the date of export of goods or completion of services and a buyer gives a promissory note or accepts a bill of exchange, ECGD may give an unconditional guarantee to the exporter's bank that it will pay 100 per cent of any sum which is three months overdue. To operate the scheme the exporter presents the bill or note to the bank concerned after shipment of the goods with the appropriate evidence and a standard form of warranty that ECGD cover for the transaction is in order. United Kingdom banks have agreed to finance 100 per cent of the principal value of such transactions charging interest at half a per cent over the minimum lending or base rate. The premium for one year is paid by the exporter in advance. He also signs a recourse undertaking giving ECGD the right to recover sums due in advance of or in excess of claims payable under the standard ECGD policy.

Comprehensive Open Account Guarantees

When trade is done on an open account, unsupported by bills or notes ECGD will give a guarantee to the financing bank on lines similar to the bills guarantee. To borrow against the guarantee the exporter provides evidence of shipment and warranty that the transaction is insured with ECGD, gives a copy invoice showing the terms of payment and also gives the bank a promissory note to cover repayment of the loan. United Kingdom banks have also agreed to finance lending against these guarantees. Should an exporter not honour a promissory note ECGD pays the bank and recovers from the exporter.

Specific Guarantees to Banks

If the terms of payment are two years or more ECGD will provide supplementary guarantees offering similar security to the financing banks, but each contract is considered. These guarantees could promise unconditional payment to the bank of 100 per cent of any bill or note against which payment has not been received by three months after the due date of payment. The banks will finance the credit in such cases, without recourse to the exporter, at a fixed rate of interest in accordance with guidelines for officially supported export credit.

Guarantees for buyer credit financing

For contracts involving substantial credit it may be more necessary for a buyer to be financed by a UK bank. This is an important feature of financing capital projects, particularly construction.

Buyer Credit Guarantees

Buyer Credit Guarantees are available to banks making such loans in respect of contracts of £1 m. or more. Under a guarantee an overseas purchaser is normally required to pay direct to a supplier up to 20 per cent of the contract price, including down-payment, on signature of the contract. The remainder is paid to the supplier from a loan made to the buyer or a bank in his country by a UK bank and guaranteed by ECGD for 100 per cent of capital and interest against non-payment. The contract may include some foreign goods and services, but

72

the amount of the loan will normally be less than the UK goods and services to be supplied.

Three legal agreements must be concluded at the same time covering: a supply contract between the UK supplier and the overseas buyer for the supply of plant and equipment and possibly for the construction of the project; a loan agreement between a UK lender and an overseas borrower to provide finance for the bulk of the payments under the supply contract or contracts; and a guarantee given by ECGD to the UK lender to cover the risk of non-payment of principal or interest. However, the only direct contractual relationship between ECDG and the exporter is the premium agreement by which the exporter agrees to pay for the ECGD guarantee of the loan which finances the contract. Although the guarantee is given to the lender it is the supplier who benefits and is accordingly responsible for payment of the premium. A guarantee can also be given for the full amount of a foreign currency loan.

The ECGD will also cover, on a buyer credit basis, loans made available by UK banks to overseas borrowers to facilitate the purchase of capital goods from UK suppliers under an overall financing scheme. The proportion of contracts to be financed from the loan, up to 85 per cent, is specified in the financial arrangements, the balance being paid on signature and before shipment direct by the buyer to the exporter. This cover and for buyer credit guarantees is only for the post-shipment part of the contract price. Cover is available selectively for pre-shipment bank finance for overseas contracts insured with ECGD on a buyer credit basis with a contract value of £1 m. or more. Ways in which a buyer or client, contractor or banker relate to each other is outlined in Fig. 9.2.

Cost-escalation, performance bonds and consortium insurance

Cost-escalation cover

Increases of costs can occur during lengthy manufacturing periods. The ECGD has powers to provide cover against UK cost increases on part of major capital goods contracts or complete projects worth at least £2 m. which have a manufacturing period of at least two years. Contracts for consultancy and other services may also be considered for this cover. When a contract is for the supply of several similar items, each must be worth £500 000 or more. The scheme is not available for contracts with EEC countries. To establish how much of the basic UK costs are eligible for cover under the scheme, ECGD usually treats as eligible a standard proportion of these costs which it is intended should be incurred within the period of cover. For cash business this proportion is 75 per cent and for credit 70 per cent. The ECGD and the exporter agree at the outset a threshold up to which increases in these eligible costs will be borne by the exporter or a buyer. The ECGD will then compensate for further increases of the eligible costs up to a maximum amount of annual increases which is also agreed at the outset.

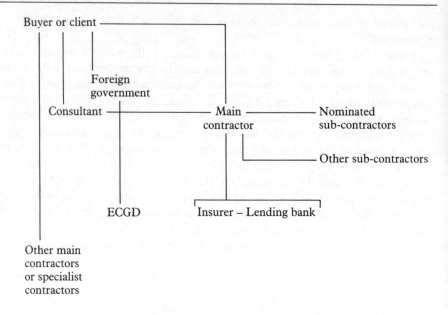

Functions of participants

Buyer or client: Seeks finance from contractor
Negotiates terms of credit
Monitors construction and certifies progress with consultant
Repays banks and others

Contractor: Seeks bank and government export agencies finance terms for a specific
project and buyer
Provides bank and buyer with information about project
Undertakes project and has payments certified and receives payments

Banker: Indicates amounts and terms of credit available to buyer
Arranges loans and credits
Disburses money against certified progress payments invoices
Collects interest and principal sum from buyer

Figure 9.2 Buyer credit. Relationships in a contract

Performance bonds

For contracts worth £1 m. or more on cash or near-cash terms which are insured
with ECGD against the normal credit risks, the ECGD will provide support for
the issue of performance bonds. The ECGD does not provide bonds, but gives
support by an indemnity to a bank or surety company which is willing to issue
the bond. Under an indemnity, the ECGD is unconditionally liable to reimburse
the bond giver in full for the amount of the bond call. Any payment by the
ECGD to a bondholder becomes subject to claim by the ECGD from the contractor involved under a related recourse agreement. The ECGD will refund the
contractor if it is established that there is no default under the terms of the con-

tract, or that failure to comply is due to specified causes outside the control of the contractor. The ECGD can also give similar support for tender, advance payment, and progress payment bonds. The premiums for the support of tender bonds is £1 per £100 per annum on the bond value, and for all other bonds it is £1.25 per £100 on the bond value. The minumum charge is for one year.

Cover against unfair calling of bonds

There is also insurance for exporters against the unfair calling of bonds raised without ECGD support. This cover is available for any contract on cash or credit terms insured under a normal ECGD guarantee, provided the form of the bond is acceptable to the ECGD and the buying country is considered suitable. Insurance takes the form of an addendum to the basic guarantee. The ECGD agrees to reimburse the exporter for 100 per cent of any loss due to the calling of a bond if it is subsequently shown that the exporter was not in default in performance of the contract, or if any failure is due to specified events outside the control of the exporter. The premium for 'unfair calling' insurance is £0.50 per £100 per annum with a minimum of one year.

Projects participants insolvency cover

Members of a consortium involved in large contracts overseas can be exposed to heavy losses which they may be unable to bear following the insolvency of a member of the consortium. The ECGD can insure main contractors or consortium members participating in major export projects of £20 m. or more for 90 per cent of any loss arising from unavoidable costs, expenses or damages due to the insolvency of a sub-contractor or consortium member. The facility is available to UK companies for joint venture or sub-contract relationships with other UK companies, or non-UK companies in some cases.

The main contractor or consortium member must nominate the amount and period for which cover is required. The premium for this cover is 1 per cent per annum on the maximum liability. As the liability of the ECGD is related to the items for which the defaulting party is contractually liable, it is necessary for sub-contractors or joint venture agreements for which insurance is required to state all additional costs which could be faced by the contractor when completing a relevant contract. Loss is established on the basis of the expenditure incurred by the policyholder when continuing the part of the project that would otherwise have been undertaken by the insolvent consortium member. Interim claims are accepted against an undertaking fron the policyholder that amounts claimed are included in the final claim lodged with the liquidator.

A valuable service

It can be seen that ECGD not only provides an extensive service, but also valuable commercial support without which many contracts would not have been possible. Few countries have such excellent support. The complexities are, nevertheless, very extensive and require detailed study in their own right and in relation to the many other aspects of establishing markets. Early discussion between ECGD and the insurer is an essential factor.

Chapter 10

Insurance for overseas projects and sales

A need for cover

Exporting of any type involves risks which are usually far greater than anything experienced with home sales or construction. This has been recognized by governments and has, in many cases, resulted in the provision of export credits insurances and guarantees services. There is in addition a need for extensive insurance of other aspects of overseas design, manufacturing, sales and construction work which must be considered when quoting prices for services and products and when preparing a tender for a contract. Too often insurance is considered late in a sequence of pricing. Early study is essential if problems and costs are to be related to a project. Insurance will influence a final price but is almost invariably a small percentage in relation to a possible loss. Even if insurance is one of the last items to be assessed, it is essential to allow a reasonable time to obtain the best terms.

A contract may stipulate particular conditions which look familiar, but on interpretation have a different meaning. The ultimate responsibility for arranging insurance must be established and excepted risks analysed. Those that have little relevance in the UK, such as for earthquakes, riots and the like, have great significance in many overseas areas. For consultants, professional indemnity will be as vital as in the UK, manufacturers will need to relate their conditions of sale to a buyer's conditions and contractors must assess the insurance conditions of the contract concerned. This may be drafted locally and be governed by specific local laws, such as Sharia law in Saudi Arabia, by modified national conditions put forward by a designer or as a result of earlier associations, such as in some African countries and in the Far East, and by the International Conditions of Contract (International) for Works of Civil Engineering Construction and those for electrical and mechanical works. In some countries decennial liability may be imposed. This is particularly likely in countries which had earlier connections with France, but others are now adopting the principle.

Buyers in some countries may be required to have insurance arranged in their own country. Nationalism may be a reason, but usually this is due to balance of payments factors or a wish to create their own insurance expertise. Some countries have state companies, often monopolies. There may be a need to use only local insurance brokers. Throughout, the services of a UK adviser will be invaluable, particularly when insurance is part of a joint venture agreement or when loss adjustment arises. Using local brokers, underwriting and other services clearly leads to language problems and, in many cases, tax difficulties. Types of cover which may require consideration are listed below.

- Advance Payment Bond. Guarantees refund of advance payments made.
- Bid Bond. Guarantees contract will be undertaken.
- Consortium Cover. Protects against losses arising through the default of any member of a consortium.
- Contractors All Risks Cover. Protects contractor against losses arising from normal hazards during a contract.
- Contract Ratification or Repudiation Cover. Insures costs incurred should signed contracts be repudiated or not ratified prior to delivery.
- Customs Bond. Replaces customs payment for goods imported which are destined to be re-exported.
- Delay Penalty Insurance. Protects against penalties following delays beyond the control of the insured.
- Employer's Liability. Protects against legal liability or bodily injury or disease to employees.
- Export Credits Guarantee. Protects against non-payment after fulfilment of contract conditions through commercial or political risks.
- Failure of Supplier/Sub-contractor. Protects against losses on default.
- Inflation Cover. Protects an exporter against the effects of inflation of a contract price.
- Insolvency of Buyer. Protects against losses arising from insolvency of a buyer.
- International Investment Insurance. Covers against loss of investment overseas.
- Maintenance Bond. Guarantees the employer against a contractor defaulting in maintenance obligations after early release of retention monies.
- Marine Insurance. Covers against risks incurred in delivery of goods overseas.
- On Demand Bond Indemnity. Protects against unfair calling of a bond where the wording requires payment on first demand.
- Performance Penalty Insurance. Insures against penalties incurred should machinery not meet specification.
- Performance/Supply Bond. Guarantees performance of the contract.
- Products Guarantee. Protects against losses arising from the failure of a product to meet specifications.
- Professional Indemnity. Protects against liability arising from errors or omissions in design, supervision, management on contract, works and machinery.
- Public Liability Cover. Indemnifies against legal liability for death, bodily injury or property damage to third parties.
- Recourse Indemnity. Protects an exporter against contingent liability arising out of recourse agreements with ECGD.
- Retention Bond. Issued to an employer in substitution for retention monies.

Professional indemnity

A consultant working overseas may be employed by a direct contract with a principal, in support of another consultant responsible for a whole project, in a joint venture or consortium with other consultants, as a designer to a contractor, or by a principal to check the designs of another. Whatever the appointment, failure to exercise reasonable care and skill may result in substantial losses to the principal, bodily injury or material damage to third parties. In the absence of provision to the contrary a consultant may be held responsible for the consequences of a failure. Most legislation recognizes such responsibility and a consultant is expected to carry out work in the relevant discipline with a reasonable standard of skill or be liable at law. Liability will normally be unlimited as far as third parties are concerned and may be unlimited between the consultant and client, but liability can be limited by the terms of an appointment. Decennial liability may also apply.

Apart from the failure or partial failure of a design, the circumstances in which a consultant may be liable for negligence are extensive. Most consultants arrange a professional indemnity insurance to cover against such liabilities, particularly in respect of design. The market for this is limited, and the extent of cover required to cater for a full claim on a large project could exceed the available market capacity leaving a consultant with a massive uninsured liability.

A consultant will usually seek to reduce potential liabilities and obtain as much cover as possible. Factors affecting a consultant's ability to identify the areas of risk depend on obtaining an adequate brief from the buyer or employer who must appreciate the need to analyse and minimise risks and who will include protection in the conditions of engagement of a consultant and on a consultant's ability to assess the design of a project and undertake detailed investigations, perhaps on a speculative basis.

Cover for a contractor

No insurance programme can provide unlimited or perfect cover, but a minimum level can be obtained. Most essential will be a contractor's all risks policy which, subject to some exceptions, covers the works erected in performance of a contract and the materials, plant and other items belonging to or the responsibility of the contractor for the purposes of a contract. During the maintenance period such cover would be limited to loss or damage to the works arising from a cause occurring prior to the maintenance period or occasioned by a contractor during the maintenance period while carrying out obligations under a contract. A general third party policy covers a contractor in respect of death or personal injury to any person other than employees of the contractor and loss of or damage to property, other than that otherwise insured, arising out of the performance of the contract, but excluding liability in respect of motor vehicles licensed for road use. The latter must be covered by a separate policy.

Cover for plant and materials to be used in the contract while in transit to a site must be covered by a marine or overland transportation policy. Storage at the site would normally be covered separately. The contractor has a statutory liability in respect of death or personal injury suffered in the course of their employment by any employee who qualifies for inclusion under locally existing compensation schemes. This is usually known as a workmen's compensation policy. A contractor may also need a group personal accident policy applicable when an employee of a contractor suffers accidental death or personal injury to provide compensation, whether or not any other person is legally liable to pay damages to the injured employee. This can relate to the employer's liability policy which covers common law liability for any death or personal injury suffered by any employee in the course of employment. The subjects covered by a contractor's all risks policy are: general exclusions; the period of cover; material damage and special exclusions; provisions applying; sums insured; basis of loss settlement; extensions of cover; third party liability and special exclusions and special conditions.

Manufactured goods

A manufacturer selling a product to a given specification may be faced with a need to insure goods in transit to the requirements of a buyer. This may be to the FOB point or CIF to the destination point. Manufacturers are increasingly required to accept a decennial liability, but this depends on acceptance on delivery by a designer and use by a contractor. For the most part the main insurance problems for a manufacturer, apart from during the course of manufacture, are demands for penalties for delays or failure to supply, failure to meet a specification, loss in transit and, increasingly, the need to provide a performance bond which follows from demands on a contractor. In all cases skilled insurance advice is essential, but the terms and conditions of a quotation for the supply of a product and those for sale are critical. Factors which must be covered in terms and conditions of sale for manufactured goods which can affect insurance planning include: the acceptance of orders; quotations; prices and terms of payment; exclusion of warranties; patents and designs; delivery; packing; damage or loss in transit; resale; *force majeure*; interpretation and the extent of conditions.

Cover by the employer

The employer will need to cover legal liability to third parties for injury, loss or damage. However, part of this liability may be covered by a contractor. The employer should have a programme of other insurance to cover liabilities which would otherwise be unprotected, for example liability for injury to employees and when any part of a project is taken over by another contractor.

Bonding

Bonds may be provided through a number of sources of which the private sector insurance market is one. Bid bonds are intended to discourage unsound bids, usually for a small percentage of a contract price. They become void when a contract has been awarded. However, if a contract is awarded to the bonded bidder, the bond only becomes void on delivery of an appropriate performance bond by the bid bond underwriters. Performance bonds can be of two types. The first gives indemnity to protect an employer in case of a seller defaulting or becoming insolvent. It becomes payable in the event of breach of contract by the seller. The second, now common in some Middle East countries, enables an employer to enforce payment by request. No breach of contract need be alleged or proved, although some bonds specify conditions as a prerequisite to payment. A retention bond enables the release of retention moneys to a seller, which can involve substantial amounts, particularly in large contracts. This type of bond usually requires the seller to carry a percentage of any loss. Normally, a supplier must provide bond underwriters with a counter-indemnity permitting recovery of any sum that must be paid under the terms of a bond.

Conditions of Contract (*International*)

With the growing use of the international conditions of contract it is necessary to appreciate the implications for insurance cover stated. The key clauses are 20–29.

Care of works

Clause 20 (1) makes a contractor responsible from commencement until date in completion certificate and for outstanding work undertaken to finish during maintenance, and liable for damage caused while carrying out obligations under clause 49 and under clause 50 for fault during progress of works or maintenance period.

Excepted risks

Clause 20 (2) exceptions are: use or occupation by employer regardless of whether or not completion certificate issued; those due to design; any such operation of forces of nature that an experienced contractor could not foresee or reasonably make provision for or insure against and explosives, radioactivity, nuclear activity, pressure waves and the like.

Insurance of works

Clause 21 covers insurance in the joint names of the contractor and employer in accordance with contract obligations to cover works and loss and damage arising during maintenance period caused by prior fault or defect. This covers works executed to estimated current contract value and materials for incorporation in works at replacement values and constructional plant to replacement value. It must cover obligations under clause 49 and clause 50

Third party liability

Clause 22 requires a contractor to indemnify the employer for injury or damage to property. Main exclusions are when caused by the employer's negligence, but a contractor's contributory negligence is taken into account.

Third party insurance

Clause 23 requires that a contractor must effect third party insurance except for cases excepted in previous clauses. The policy must indemnify an employer in any case when a contractor could claim indemnity under the policy.

Liability for accident or injury to workmen

Clause 24 (1). A contractor must indemnify an employer in respect of such claims, except when arising through an employer's negligence.

Insurance against accident or injury to workmen

Clause 24 (2). The contractor must effect insurance.

Interference with traffic and adjoining premises

Clause 29. When a contractor is liable for such claims the employer must be covered.

Table 10.1 details risks and shows where there is a loss or a potential loss, liability or responsibility for an employer, engineer and contractor. In some cases government compensation may be given for losses not otherwise covered as in the case of war risks. Liabilities due to nuclear hazards, damage to goods in transit, faulty goods and unforeseen risks may be negated by recourse against those responsible, but much depends on the factors involved in the legal conditions involved.

Foresight and pessimism

The need for early, skilled and detailed professional guidance can not be underestimated. Those responsible for pursuing overseas construction markets or managing design, manufacturing or a project must be able to assess the need for

Table 10.1 Risks and responsibilities

Risk	Employer	Engineer	Contractor
During project or maintenance period			
War, riot, commotion, disorder	×		
Nuclear and pressurewave risks	×		
Hazardous explosives	×		
Unforeseeable forces of nature	×		
Loss or damage during transit	×		×
Faulty materials or workmanship	×		×
Due to engineer's negligent design	×	×	
Due to engineer's non-negligent design	×		
Use or occupation by the employer	×		
All other causes	×		
Loss or damage to plant			
War, riot, commotion, disorder	×		×
Nuclear and pressurewave risks	×		×
Loss or damage during transit			×
All other causes			×
Third party losses			
Unavoidable result of carrying out contract	×		
Negligence of employer	×		
Negligence of contractor			×
Professional negligence of engineer		×	
Other negligence of engineer		×	
Injury to employees of contractor/sub-contractor			
Negligence of contractor			×
Negligence of employer	×		
Professional negligence of engineer		×	
Other negligence of engineer		×	

insurance and the implications of lack of cover or insufficient cover. In all cases foresight and imagination are necessary to analyse risks, with the support of a professional adviser who has the specialized knowledge of particular types of business. In addition, a degree of pessimism is, regrettably, likely to be a help rather than a hindrance.

Chapter 11

Taxation

Profit motive

Governments may encourage trade for political, diplomatic, financial, employment and a variety of other reasons of which profits or a contribution to a balance of payments are only part. Commercial undertakings have no such spread of interest, their primary objective is immediate or future profit. For these reasons, among others, taxation factors are an essential part of pursuing overseas activities. It may be pointless to pursue an activity if taxation is likely to absorb an unreasonable part of income. Taxation must be related to that at home and other financial commitments. It is also a subject that requires skilled and early professional advice. As a part of financial planning, it will affect the type of trading structure used, accounting methods, profits and cash flow. These in turn will be governed by local laws, taxation allowances, withholding taxes, double taxation relief and personal taxation which can have considerable influence on recruitment and, therefore, on the conduct and success of an undertaking.

While guidance should be obtained from skilled specialists, those responsible for obtaining, directing or managing overseas work should appreciate what is involved and the impact on planning. This should assess specifically what taxes would be due on profits earned, taxes due on profits directed to the UK and when actually repatriated. Throughout, it is essential to minimize the still relatively high UK rate of taxation. The key legislation comprises the Taxes Acts, Income and Corporation Taxes Act, 1970 and Finance Acts.

Complex home planning, tax avoidance schemes, use of tax havens and investing methods may be pointless if home and local taxation is likely to be similar. This is important if trading conditions in the country concerned make it essential to repatriate profits. Remember, also, that local tax avoidance can influence obtaining work in the future and be construed as disinterest in the future of the country concerned. This should not be underestimated. As in so many matters foresight and judgement are needed, but must also be related to the return on the investment being made, risks involved and contribution to home profits, and to the fact that some taxation will almost invariably be paid somewhere.

Local structure and double taxation

Much will depend on what use is made of a local partnership or company and whether or not the activity is financed by a substantial UK public company or by a private company. If the latter is a close company the taxation of individuals is likely to be as rigorous as for private individuals or a professional partnership. While exchange control restraints, which were removed in 1979 and resulted in about £10.7 bn. being invested internationally in 1981 compared with approximately £6.5 bn. in 1979, no longer apply, there are still many restrictions which make tax avoidance difficult, not least on the use of tax havens and the introduction of an overseas company between a UK company and another overseas trading subsidiary.

Companies resident in the UK will face problems of double taxation. These apply to investments and other operations. The company must know whether or not it will be subject to overseas tax, how to minimize the tax and obtain relief against taxes. The decision to operate through an overseas subsidiary company or through an overseas branch of a UK company is, therefore, critical, and needs an understanding of double taxation and the availability of relief as a basis for such decisions. Companies resident in the UK may not transfer an existing business to the control of a non-resident without first obtaining consent from the Treasury. In most cases the entitlement of a country to tax a company depends on whether or not it is resident or has business in the country as opposed to with the country, but it also depends on receiving income from a source situated there. The UK has a number of double tax treaties with other countries. These regulate the extent and entitlement to tax classes of profits or gains and attempt to harmonize laws so that a common interpretation may be obtained.

A company may be taxable in an overseas country even though not resident there. To determine whether or not a company is trading in an overseas country or with the country depends on the local law. If this imposes tax there may be recourse to a treaty definition limiting the entitlement of the overseas country to impose tax. The definition of a permanent establishment differs with treaties, but generally means a place of management, office, factory, mine, quarry or construction project which exists for more than a year. Facilities for storage or for purchasing do not necessarily imply a permanent establishment. However, if there is a representative, who is not an agent or broker, in an overseas country who exercises authority to conclude contracts in the name of a company it may be treated as having a permanent establishment there.

Rules for interpreting residence for tax purposes also differ. It may be determined by the place of incorporation. United Kingdom law considers a company to be resident in a country where central management and control are exercised. Control is always a matter of fact, but the following are indications of the criteria of residence: location of meetings of directors; composition of the board of directors; location of office; location of accounts; extent to which a board controls and is responsible for trading and finance; control by local employees;

direction from a board located elsewhere and the location of meetings of shareholders.

Double taxation relief

Agreements are either comprehensive or limited to deal with the profits of shipping and air transport. There are over seventy of the former, but only a few of the latter. They are governed by Taxes Acts. Rates of UK tax are not usually less than those of similar taxes overseas, and, while credit relief would normally be obtained for such overseas taxes under unilateral provisions, bilateral treaties give benefits to a UK taxpayer. Treaties may reduce the scope or rates of tax in an overseas country so that a UK taxpayer may defer payment of tax until the UK corporation tax payment date, whereas without such protection tax may be payable earlier in an overseas country.

When it is possible to accumulate income or gains overseas a treaty can be used to ensure that the maximum amount of post-tax profits are directed to suitable locations, and repatriation requirements may be satisfied with the repatriation level of all overseas subsidiaries thus enabling retentions to be made of certain profits at the expense of others. There may be situations when obtaining credit relief in the UK for overseas taxes is prevented. In these circumstances treaty provisions for reduced rates of foreign tax may help. Exposure to overseas tax under local laws may be reduced or eliminated as a result of a treaty, but, while a treaty may remove some uncertainties, there may be differences of interpretation by two contracting countries or states. The structure of a typical comprehensive treaty relating to taxes on income and capital is:

Scope of agreement

Taxes covered and effective dates.

Definitions

General, residence and permanent establishment.

Taxation of income

Property; commercial and industrial profits; shipping and air transport; void conditions and dividends.

Taxation of capital

Reliefs and credits

Personal and elimination of double taxation.

86

Special provisions

Non-discrimination; mutual agreements; exchange of information; diplomatic privileges and territorial extension.

Protocol

Starting date; reliefs under any previous agreement and procedure for termination.

Tax relief

Before establishing a complex tax avoidance structure it is wise to clarify what tax relief is available in the UK and elsewhere. Some countries give far greater relief than the UK. Tax relief may be available as a result of a bilateral taxation treaty, or unilaterally. A treaty may not prevent profits being taxed twice, but will generally provide for some relief from UK taxation. This may be by offsetting overseas tax against UK tax or by deducting overseas tax from UK taxable profits. Overseas tax costs may also be added to UK losses and be carried forward. For this reason, among others, overseas taxation must be a part of a total financial assessment. Capital allowances, stock relief and past losses may be involved, but if there is no liability obtaining relief for overseas taxes may be difficult, and it may be necessary to delay or divert remittances, obtain relief by deduction or establish postponement.

Unilateral relief provisions, which apply only when treaty relief is not available establish a right to obtain credit relief. If it is an advantage to claim relief by deduction instead of credit relief this may be done. Entitlement to unilateral credit relief depends on overseas tax being admissible, related to income or gains arising in an overseas taxing country or state and, in the case of dividend income, the overseas tax must be one which would not have been borne if the dividend had not been paid. The Board of Inland Revenue publishes details of overseas taxes which are admissible. It represents a view, but has no legal status, and does not preclude appeal by the taxpayer. Taxes payable under the law of another country or state, or levied by local authority, may also be admissible provided they correspond to UK income, corporation or capital gains taxes. As dividends are paid out of taxed income relief may be obtained for taxes relating to a dividend and for taxes indirectly related. This means that, in addition to a withholding tax on a dividend, taxes on underlying profits out of which dividends are paid are available for relief. The circumstances under which relief for underlying taxes may be obtained by a UK company are when it controls directly or indirectly not less than 10 per cent of the votes in a company paying a dividend or if it is a subsidiary of a company which has such control.

Credit relief

There are rules which determine when credit relief is available. Known as the credit code, the rules apply to unilateral credit relief and to treaty credit relief. The main principles are that credit relief will not be given to companies which are not resident in the UK; when a taxpayer is entitled to credit relief, and has not elected against credit relief in favour of relief by deduction, the amount to be included in a UK taxable profit is the gross amount of overseas income before deduction of overseas taxes; and the amount of a credit for overseas tax allowed against UK corporation tax cannot exceed the amount of corporation tax which, but for credit relief, would have been payable on the gross overseas income.

Tax on profits of an overseas subsidiary company or branch

Profit of overseas resident subsidiary companies are not taxable in the UK until dividends are declared payable. Structuring overseas investment through a subsidiary allows flexibility of dividend payments to the UK. Where minority interests could prevent flexibility it may be necessary to introduce an overseas holding company as the trading company. Overseas branch profits of a UK company are, however, taxable in the UK in the relevant accounting period. This may result in the loss of some double tax relief if there are UK tax losses, and the company should ensure that overseas branch profits arise in an associated UK company which has no tax losses, transfer the branch business into an overseas resident subsidiary company or elect for relief by deduction. When credit relief is restricted in respect of overseas tax on the trading operations of an overseas branch because UK capital allowances exceed the corresponding overseas allowances, the company may elect to postpone utilization of part of the UK capital allowances. While postponement of allowances is permitted only in these circumstances a company may elect to waive part or all of the first-year capital allowances to reduce tax losses which might otherwise have restricted relief. Allowances waived may be obtained in subsequent years at a rate of 25 per cent of the reducing balance.

Tax sparing relief

To encourage economic and social progress in some developing treaty countries which offer tax holidays UK law allows in principle exempted overseas taxes to be deemed to have been paid, and as such be available to eliminate liabilities to UK corporation tax. This is called tax sparing and allows overseas tax incentives to be kept intact by UK companies which, under the credit code, would otherwise have lost the benefit of such incentives. It can be a valuable tax advantage to those involved in construction.

Offshore financial centres

Establishing an overseas subsidiary or holding company to avoid or delay tax requires the most skilled advice, not only from accountants and lawyers. All involved in directing and managing an undertaking may need to be consulted as the use of a subsidiary or offshore financial centre, also known as a tax haven, may be dictated by trading factors, but can also influence trading patterns or planning.

An overseas subsidiary company may be located in the country where trading is being carried out, or an intermediate company may be created. For public companies there are already a number of methods established by locating, for example, in the Channel Islands, Cyprus, Luxembourg, the Netherlands and the Netherlands Antilles, Switzerland and a variety of others, each with advantages and disadvantages. Under such an arrangement a UK company may establish a holding company in, for example, the Channel Islands. The company could make investments overseas elsewhere, possibly where profits are not taxed or, if tax is in principle payable, it is not collected or taxed at a low rate. In these circumstances the profits would accrue to the company with little or no loss. On reaching the company repatriation requirements could be met without the profits incurring UK corporation tax as they would be outside the UK for such tax purposes. This type of structure is particularly suited to public companies where there is little risk of challenge by the Board of Inland Revenue.

In the case of a smaller UK public company or one which is owned by a few shareholders there could be an assessment for tax on the profits of such a company. For this reason it may be preferable to establish an overseas holding company in, say, the Netherlands, particularly if other European or worldwide interests exist as by such arrangements the profits of overseas subsidiaries can be paid by way of dividends to that country and suffer no tax. If dividends are then paid to the UK the total rate of overseas taxes will have effectively been pooled so that the underlying rate may be reasonably high, and little UK tax will be due on the dividend received.

The UK government has become increasingly concerned about the growth of the use of tax havens and early in 1981 the Board of Inland Revenue put forward proposals aimed at preventing companies using overseas subsidiaries to avoid UK tax. They relate to UK companies deemed to be avoiding tax through overseas subsidiaries on overseas income and capital gains.

Corporate shareholders resident in the UK could be treated as liable for a pro rata share of a UK tax charge on the income and gains of overseas companies which are controlled by UK residents. The key challenge so far against this form of corporate tax avoidance has been section 482 of the Income and Corporation Taxes Act, 1970, which prevents companies from transferring their place of residence for tax purposes and prohibits certain other transactions relating to overseas companies without Treasury consent. However, following the abolition of exchange controls in 1979 it could be argued that those provisions are no longer appropriate as new legislation is needed. Repeal of Section 482 could increase the risk of loss by tax avoidance.

Control of a company could be defined by existing law and the charge apply only to those with a direct or indirect interest of 10 per cent or more in an overseas company. There would be safeguards to exclude groups involved in genuine trading and commercial activities. The charge could only arise when a company pays no tax or at a rate significantly less than in the UK. Such powers could change rules defining company residence. This is an example of the constant attempts to change legislation which can seriously affect planning and pricing. It is by no means restricted to the UK.

There are other reasons why tax havens should not be used indiscriminately. Those which have no treaty protection could incur taxes. To interpose a company in a tax haven between an operation in another country and a UK parent company achieves nothing if there is no double tax treaty and allows dividends to be subject to a withholding tax. This could increase the overall rate of overseas taxes way above the limit payable in the UK. A treaty could provide for a lower withholding tax on dividends and, therefore, be a cheaper way to repatriate profits. Remember also that tax havens also have taxes and there will certainly be legal, local directors and other expenses.

In the case of professional partnerships which are taxed in full on worldwide income some relief has been given under which a UK resident's share of an overseas partnership is exempt from part of profits. However, by using treaty arrangements it is possible for a UK partner of a selected tax haven resident partnership to be protected so that profits cannot be taxed in the UK. This involves establishing a physical presence in the relevant tax haven, thereby incurring costs, but there can also be significant tax savings.

Personal taxation

There are limited tax incentives available for a UK resident who works overseas. No longer are there tax penalties when travel and accommodation costs are borne by a company. When UK employment involves overseas work of more than thirty days deduction is allowed on 25 per cent of foreign earnings. Overseas work for parts of two fiscal years for a continuous period of 365 days can also result in deductions, but careful tax conditions are imposed. In the case of absence overseas for a complete fiscal year no tax is due, but a wife's income and capital gains are still liable even when she accompanies her husband. Employment with an overseas employer and payment for duties overseas leads to deductions of 25 per cent irrespective of the length of the employment. In all cases the definitions of domicile, residence or ordinarily resident, whether or not employment is with a UK or foreign company and the period of employment stated in a contract of employment, govern tax exemptions and tax-free periods permitted for return to UK. Assessment may depend on interpretation by a tax inspector. Again, the need for expert and up-to-date advice is very evident.

Most aspects of construction are still labour intensive. When operating overseas it is likely that technical and other manpower from the UK and elsewhere will be needed, and an overseas employment company may be established for this which will affect UK personal taxation rules and those of other countries. Some may escape tax on overseas earnings provided that some tax is paid on earnings. For this reason an employment company in a country where there is a low rate of income tax which will satisfy such a requirement and enable exemption to be obtained is preferable and is no disadvantage to a UK resident as the UK resident will pay the relevant tax, but will have it deducted from the UK tax.

Financing construction overseas

Taxation of profits, personal income and dividends and the resulting strategy while essential should not obscure other aspects of tax relating to overseas construction. Whether financing manufacturing, a project or other activity the relative amount of capital needed grows with each new activity. Careful consideration must, therefore, be given to the tax implications of financing overseas projects. One of the most important aspects is that tax relief should be obtained for any interest paid on borrowings. Ideally the best position is achieved when interest paid by a company is matched by tax relief. There is little advantage in arranging for interest to be paid out in a low tax area unless this cost can be claimed as a deduction against profits in a high tax area.

Increased borrowing power can be derived from grouping overseas subsidiaries in an overseas holding company, and interest borne by subsidiaries can be fully deductible, but must also be matched. There can be taxation and reliefs changes as a result of currency exchange rate fluctuations.

Many companies with overseas interests use an off-shore base, but it is essential to select a country where the tax system has a history of stability and is unlikely to change, particularly in providing advantages; For example so that dividends received by a company from overseas investments, provided they are not portfolio investments, are tax free, and if disposed of by subsidiaries capital gains on those disposals are not taxed. If there is exchange control it must not be restrictive. The country must also provide a comprehensive system of double tax treaties to ensure that dividends from subsidiaries suffer the lowest possible rates of withholding tax. When tax exemption does not apply to profits from countries where there are not taxes on profit it may be necessary to establish a subsidiary elsewhere so that profits can be put through a company there and suffer a minimal rate of tax. Ensure also that any base for a holding company is not deceptive. Some may not permit dividends to be accumulated tax free. Avoid, if possible, those with high administration costs. Selection must depend on conditions existing and anticipated. Many countries have had a remarkable stability of conditions, but there can be no alternative to careful selection based on expert guidance.

Planning factors

In conclusion, the key factors are to ensure that profits are earned in the lowest tax area possible. Always prepare or obtain a repatriation plan using the best route for channelling profits. By forward planning maximize and preserve tax relief on initial losses on overseas operations. Also obtain the best balance of profits and losses on exchange fluctuations and tax relief for interest on overseas borrowing.

Chapter 12

Contract documentation

A need for precision

The end product of construction, whether it be a house, flat, hospital, power station or any one of a multitude of structures or civil engineering activities will usually have been based on painstaking research into economic, business and design facts. Nevertheless, any part of a construction activity is likely to be disrupted by unknowns, and it is essential, therefore, that the design of the project and the contracts governing its execution should be as precise as possible.

In the first instance, those involved in overseas work should study contracts applicable at home and which relate to design, supply, erection and associated activities such as management of property. Some will have application overseas, and may have influenced laws drafted in other countries. In some countries rigid contract laws based on local needs may be the only ones relevant or acceptable. The hazards in such cases can only be overcome by experience and by the use of advice from specialists. For the most part, however, Conditions of Contract (International) for Works of Civil Engineering Construction and those for electrical and mechanical works may be expected to have a large influence. Each can govern the form of tender and the subsequent agreements, and if not accepted in full by an employer or consultant, or, indeed, a contractor, can form a basis for a contract. The supply and shipping contracts will be as important, but are usually related to the avtivities of each organization involved, whereas the international conditions of contract have widespread international acceptance by all sectors of international construction, particularly the Fédération Internationale des Ingénieurs-Conseils whose initials FIDIC are generally used to describe the conditions. They generally have widespread support by governments and other users. A full list of construction industries organizations that approve the conditions is given below.

- Fédération Internationale des Ingénieurs-Conseils.
- Fédération Internationale Européene de la Construction.
- International Federation of Asian and Western Pacific Contractors Associations.
- La Federacion Interamericana de la Industria de la Construccion.
- The Associated General Contractors of America.

Basis of a contract

Preparation of a contract, its signature and interpretation are matters which require specialists' skills and advice, and about which much has been said and written. They are, nevertheless, means to an end and not ends in themselves, and for these reasons must be a part of management, marketing and decision-taking generally. Their study should not be avoided, if only to realize the cost and complications arising from recourse to lawyers, arbitration and courts when management decisions could have avoided this.

Contract conditions should provide a method of working and for overcoming problems, usually by those involved in the management of design, supply and erection rather than in legal matters. Any contract should be influenced by this. Conditions should be equitable, as precise as possible and objectives must be clear and capable of interpretation by users. This applies particularly to the main contract such as FIDIC as supply and other factors are likely to stem from this rather than the reverse. However, supply contracts will often be based on the standard conditions for a supplier, but there may be penalties agreed for late supply, and, in some cases, a contractor may pass on bond liabilities.

There are always likely to be differences of interpretation of any contract, and when different countries, languages, customs and laws are involved even an international agreed standard form of contract cannot overcome every problem. Too often it is the contractor and supplier who suffer as a result of their capital investment and relevant lack of flexibility compared with that of an employer, consultant or carrier. There must, therefore, be careful study of changed conditions or the qualification of clauses even when they are inducements to the early award of a contract. Reversal of a change after signature is usually impossible on practical grounds or because the legal barriers and time scale may be insurmountable and, in addition, because the local methods of business may render any change unacceptable even though it may be seen as reasonable and logical. The structures of the current FIDIC conditions are summarized below. Each of the clauses outlined will be influenced by the many factors that govern the analysis of the location of a project and the project itself. In the contract each party will seek to protect individual interests.

Civil engineering construction

Part I. General conditions
Definitions and interpretation; engineer and engineer's representative; assignment and sub-letting; contract documents; general obligations; labour; materials and workmanship; commencement time and delays; maintenance and defects; alterations, additions and omissions; plant, temporary works and materials; measurement; provisional sums; nominated sub-contractors; certificates and payments; remedies and powers; special risks; frustration; settlement of disputes; notices; default of employer; changes of costs and legislation and currency and rates of exchange.

Part II. Conditions of particular application
These depend on a particular contract, but would include such factors as definitions, language, law, bonds, insurance, manpower, sub-contractors, materials, plant, payments, completion, bonuses, damages, maintenance and amendments.

Part III. Conditions of particular application to dredging and reclamation work
Form of tender

Form of agreement

Electrical and mechanical works

Part I. General conditions
Definitions and interpretation; engineer's supervision, assignment and subletting; extent of contract; contract documents; general obligations; labour; workmanship and materials; taking-over and defects; variations and omissions; ownership of plant; ownership of contractor's equipment; certificates and payment; remedies and power; outbreak of war; frustration; settlement of disputes and arbitration; notices; default of employer; variation of costs and customs and import duties.

Part II. Conditions of particular application
These would generally follow those for civil engineering construction, but applied to electrical and mechanical works.

Form of tender

Form of agreement

Prequalification and tender

Before contract conditions have been studied in any great detail it may have been necessary for a contractor to have pre-qualified for a project. This is essentially the presentation of evidence of ability to undertake the type and scale of project involved. Too often only a limited time is available for this. Time is also often short for preparing a tender. This may also require the presentation of a bid bond. Other critical factors in a tender are third party insurance, start of the project, completion date, liquidated damages, bonus, maintenance period, adjustment of provisional sums, retentions and interim certificates which will stipulate precise timings, percentages or amounts, or none as the case may be.

No tender will be possible without proper designs, including complete site plans, levels, drawings, specifications, and a bill of materials or schedules. These may have been prepared by a designer from any part of the world, possibly one unfamiliar with the practices in a contractor's country, and, indeed, by someone with little international experience. In addition the design may have been prepared much earlier and be based on practices, materials and products that are obsolete or at best obsolescent. Some contractors may bid on the basis of changes being acceptable and may base their price on likely advantages to be gained. The consultant will be the decision-maker in this case, but remember it is the employer who pays the contractor and is able to terminate a contract in the event of a contractor failing to perform or due to *force majeure*.

Failure to perform may be a matter of fact or of opinion. For this reason a contractor must always assess the ability to work under existing economic, social, climatic and other conditions which may be divorced from actual skills and experience. Rules for arbitration are, therefore, likely to be a matter for very careful study. Disputes may be settled under Rules of Conciliation and Arbitration of the International Chamber of Commerce, or someone appointed under the rules. Settlement may, however, only be permitted as a result of arbitration in the country of the employer and under local laws. Clearly this could be hazardous, time consuming and expensive. Should it be necessary to resolve a claim there are a number of factors to be considered such as:

Basic information

Project and risks during performance and management.

Contract

Basis; documents; language and law; payment provisions; disputes clauses and other factors.

Variations

Supply quality; adverse site conditions and pricing variations.

Delays

Delayed and suspended performance; performance pressures and pricing.

Preparation of a claim

Proof; liability; cause and damages.

Negotiation

Preparation; timing and method.

Arbitration

Preliminary factors; arbitration basis; arbitration with governments; contract review; recognition of a claim; schedules and cost control methods.

Management contracts

For a variety of reasons a construction contractor or manufacturer may accept, or, indeed, prefer, to work only on a basis of a management contract. This could be a condition of supply of some products, particularly when assembly skills are needed and when equipment is being used for a first time. For contractors the decision may be based on accepting a smaller profit and risk than is involved on a direct main contract or sub-contract, as a means of gaining experience in a country or because management and other skills within the organization concerned limit involvement. In some cases a local contractor may have been able to obtain a contract on the financial strength available, but dependent upon importing management. Management contracts are a growing feature of international construction. Planning the application of such a method of working cannot be separated from all aspects of contract documentation nor from the working relationships of main and specialist contractors and sub-contractors. It may also be an important feature of a joint venture agreement.

Suppliers' contracts

A supplier may be asked to provide a performance bond and to agree a specific supply programme. In the absence of these the terms and conditions of sale of a supplier or manufacturer will generally stipulate that supply is based on terms and conditions which supersede and exclude all general or special terms or conditions imposed or sought by a customer at any time in relation to any orders. Typical terms and conditions of sale are: the basis of acceptance; quotations; prices and terms of payment; exclusion of warranties; patents and designs; delivery; packing; damage or loss in transit; resale; *force majeure*; interpretation; extent of conditions and law and language.

Acceptance of conditions by a supplier will normally be assumed unless told otherwise. Any variation would have to be agreed in writing. A quotation does not constitute an offer to supply goods or carry out work. Usually this would be agreed in writing after acceptance of a quotation, but would depend on availability of goods and materials necessary and any period of validity given for price. The latter could be influenced by indexing or another formula. When supply is to be spread over a lengthy period something of this nature would be essential. Prices quoted would be net. Any discounts given may be stated separately and depend on the size of an order, delivery schedule and point of delivery. Payments could be due to an agent even when a quotation and order are negotiated directly between the supplier and customer. In the absence of payment for any part of goods ordered following a buyer's insolvency a supplier may have a general right of lien on all goods held, even though payment may have been received for them, for the unpaid price of other goods sold and delivered to the buyer under the same contract.

A condition of warranty may be sought by a customer. Supply may be to a

97

recognized manufacturing specification, but exclusions will almost certainly be stated, particularly liabilities for consequential loss and infringement of patents and designs. Only approximate delivery dates will be agreed and delays will not permit cancellation or refusal by the buyer. Packing quality and costs conditions must be stated, so, too, must liabilities and exclusions for damage and loss in transit. Purchase free on board (FOB), cost and freight (C and F) and cost, insurance and freight (CIF) will normally allocate liabilities for loss. Should resale be necessary by the buyer this is likely to be permitted only with existing trade marks and under conditions agreed for the first sale. *Force majeure* conditions and interpretation of the terms and conditions of sale may be expected to provide for circumstances not otherwise covered. Supply may also be based on a licensing or know-how agreement.

Freight contracts

Freight, whether carried by sea, land or air will be subject to a carrier's conditions. Services will be provided on stipulated terms and customers entering into a transaction will be expected to warrant that they are owners of the goods or agents of the owners and authorized to act on their behalf. Pending forwarding or delivery the costs involved will be passed to customers, who will also be expected to allow the carrier freedom to select or change routes unless instructed to the contrary. Packing is the responsibility of the customer, but most forwarders will provide this service. They will also process documents required for arranging payment. This is a particularly important and exacting task when letters of credit are involved and when such payment documentation states specific conditions and presentation, particularly, of supporting documents, which is the case when dealing with some countries whose trading conditions are influenced by political factors. Essential matters likely to be covered in carriers conditions' are: the basis of acceptance; ownership of goods or agent; storage, handling and transportation; mode of transport; packing; time for acceptance of quotation; insurance; currency fluctuations; accuracy of descriptions; liability for duties, taxes, imposts, levies, deposits and the like; liabilities in the case of non-delivery, non-compliance and for consequential loss; discharges from liabilities; perishable goods; hazardous goods; goods of particularly high value; recovery and claims; particular and general lien; indemnities and law and language.

Inter-related skills and agreements

Construction work of all endeavours seems to involve the greatest number of inter-related skills. Not surprisingly, therefore, the matters on which agreement is needed before a project starts, during work and afterwards are extensive and complex. Carefully analysed and structured agreements and contracts are necessary, but ultimately these depend on goodwill and commonsense interpretation. Those responsible for their preparation should be consulted at the early stages of planning and should be kept informed in every possible way throughout if the best answers are to be obtained.

Chapter 13

Export services and sources of information

Travel involves time and costs

There is a well-known maxim which says that time spent in reconnaissance is never wasted. Some have an alternative adage saying that time allocated to surveys is never recovered. However, for those involved in construction, more than any other industry, site visits are essential, and start with the local assessment of every financial, commercial, industrial, social and political factor. But avoid rushing into such investigations before an objective has been determined, even though this may be fact-finding rather than agreeing a business course, and certainly before use has been made of the many home sources of information. In most developed countries, and many others, they are extensive, and in the UK with its long history of overseas trading and investment the sources are excellent. Such information may be available from government offices, commercial and industrial undertakings, with which there may already be trade, professional and trade bodies, libraries and individuals. Some may charge for information, but many offer free services as a part of public relations and promotion. Even competitors may be helpful when enquiries do not immediately impinge on their activities and when pursuit of a new and nationally important project or market is involved.

To obtain information it may be necessary or convenient to use an external specialist. This can often be more expensive than using internal staff who will know in more detail the objectives and structure of their organization, they will also retain the knowledge for future if not immediate use and also may be less costly. However, existing staff could be less objective and not have the experience or contacts required. It may also be necessary to keep such research confidential. In some cases new staff may be needed. Use of external services may, therefore, be preferable. In all cases selection of an external consultant should be made with great care as unless they have an understanding of construction their conclusions may be suspect, and certainly may have taken more time, and, therefore, expense to achieve. It would be all too easy to suggest a number of addresses of those likely to provide export services and of sources of information, but these change and it is the principles involved which are important, whether used in the UK or elsewhere.

100

Government services

Few countries can call on such widespread and comprehensive services as those provided by the UK government. The Foreign and Commonwealth Office (FCO), apart from its diplomatic role, has developed support for trade to a remarkable degree. Enquiries may be made at home, but equally important are the reports that can be obtained from diplomatic and commercial staff at embassies and high commissions overseas before a vist there may be made. But remember they are dealing with many visitors and interests. Few are likely to be specialists in one profession or commercial or industrial activity, despite probable earlier training before entering the FCO. The selection of questions is, therefore, essential and will enable use to be made of FCO contacts elsewhere. Never expect the staff involved to prepare and answer questions.

The FCO works closely with the Department of Industry and the Department of Trade. It also incorporates responsibility for support for overseas development which can often involve funds for construction or projects requiring construction support, particularly through the Overseas Development Administration (ODA). The Department of Trade has Commercial Relations and Exports Divisons covering countries and regions with valuable information and contacts. All work with the Export Services Branch, which, apart from contacts, information and statistical services, also publishes a daily export intelligence bulletin. The work of all departments is helped by area advisory groups whose chairmen and members are directly involved in trading with the countries concerned through commercial, industrial, professional or government activities. They in turn work with the British Overseas Trade Board (BOTB) which encourages exports and advises the government on policy and on trade promotion. The latter may be in the form of inward or outward missions, visitors to the UK, overseas trade fairs and exhibitions and any one of many similar activities. Detailed planning and supervision is done by the Export Services and Promotions Branch of the Department of Trade. In many cases financial support for exporting can be obtained through these sources, particularly for missions, exhibitions, market research, market entry, through a Market Entry Guarantee Scheme and, very importantly for those in the construction industry, there can be financial support for designers, manufacturers, suppliers and others pursuing major projects. The latter can be in the form of support for tendering costs, but there are conditions. This activity is clearly an important one and is handled by the Projects and Export Policy Division.

The Projects and Export Policy Division of the Department of Trade has a particularly important role in acting as a focal point for coordinating goverment support for industries pursuing capital projects overseas. It also advises on policy, particularly on aid, provides services for consultants and contractors and gives information about projects financed by international lending agencies.

In the Department of the Environment (DoE) there is a construction industry directorate responsible within government for sponsorship of the construction industry. It has links with trade and professional associations to promote interests abroad. The Construction Exports section of the Building Research Establishment (BRE), also a part of the DoE, acquires and provides technical information on overseas construction and is able to give advice and support to UK firms seeking and carrying out work overseas. Health Building Overseas, which is a part of the Department of Health and Social Security (DHSS), provides support for UK consultants and contractors seeking building contracts for health authorities overseas. It has access to medical and nursing planners. Other government departments, such as the Department of Education and Science (DES), may be expected to give advice on construction matters likely to encourage exports which are within their experience.

Manufacturers may need information on tariffs. This can be obtained from the Overseas Tariffs and Regulations Section (OTAR), which will also advise on packing, shipping and on restrictions. Remember that some countries may have special restrictions, particularly involving trade links with other countries. There may also be strategic limitations. Such matters require careful discussions with trade advisers and with governments' departments.

Although offices in London are often primary sources of information and policy, most government departments concerned with exporting have regional offices in the larger towns. Government department links with the Commission of the EEC involving such matters as financial support for African, Caribbean and Pacific (ACP) countries through the European Development Fund (EDF) can be helpful, but so, too, is direct contact with offices in London and Brussels, and with the London office of the World Bank. Contact with offices of foreign and commonwealth governments in London may be helpful, but their primary function is to encourage their own trade and much will depend on the value placed on a UK enquiry.

Commercial and industrial

An early contact for information should be the joint stock or merchant bank with which all in the construction industry are likely to have business links to some degree. Most joint stock banks have international companies with overseas offices and correspondents whose primary function is to increase trade at home and overseas. The same applies to merchant banks. Both are likely to have been involved in financing operations for every conceivable business venture, in many cases spread over decades. Their experience and grasp of local knowledge is likely to be excellent, particularly when related to the different, but complementary, knowledge that may be obtained from British embassies and high commissions.

Consultants, manufacturers and contractors can benefit from the experiences of each other. Each is likely to have existing business relationships at home and overseas which provide a commercial reason for giving advice and assistance. Even competitors can be surprisingly helpful to each other, within limitations, but do not expect too much and always prepare questions, and be prepared for rebuff when there is a conflict of interests. This is unlikely to be a personal matter. Those concerned with freight, such as shipping, airlines, haulage contractors and forwarders may have existing or past links with the country or area of interest. Even if they have not, their knowledge of complications that can arise is invaluable. They, too, have a vested interest in helping a potential customer, which a manufacturer or contractor interested or involved in overseas construction could become. Insurers, insurance brokers, commodity brokers, stockbrokers, newspapers and technical journals may also be a source of information.

Professional and trade bodies

Leading employers' bodies such as the Confederation of British Industry (CBI) often have departments specializing in fostering exports and with the legal and other complications that arise. Chambers of commerce have similar expertise, usually influenced by the industries and areas which they serve. Some may be industry orientated, such as in Birmingham, some commercially orientated, but on a very broad base, such as the London Chamber of Commerce and Industry. Most work together and have links with similar bodies overseas, some of whom may have UK offices. The Royal British Institute of British Architects (RIBA), Royal Institution of Chartered Surveyors (RICS) and other professional bodies, such as The Association of Consulting Engineers (ACE), the Institution of Civil Engineers (ICE) and those dealing with electrical engineering, mechanical and structural engineering and commercial matters, including accountancy, banking, insurance, procurement and taxation give support to their members for overseas activities, so, too, does the Federation of Civil Engineering Contractors (FCEC), National Federation of Building Trades Employers (NFBTE) and the National Council of Building Material Producers (NCBMP). But more specialized help may be obtained from the British Consultants Bureau (BCB), the Building Materials Export Group (BMEG) and The Export Group for the Constructional Industries (EGCI). The latter has particularly good contacts with international contractors' associations and federations. There are, nevertheless, numerous other professional and trade bodies concerned with professions, industries and commercial functions which may help. Their selection for approach must be a part of planning, as, too, may be contact with trade unions and the Trades Union Congress office. All will have something to offer.

Professional and trade bodies, universities and schools may offer information about technical matters affecting design and selection of products, as may research and testing laboratories concerned with particular materials such as aluminium, cement, plastics, steel and timber. The British Standards Institution also advises on foreign standards and codes of practice. It has a special unit, Technical Help to Exporters (THE), concerned with this, but also has widespread contacts with overseas standards-making bodies and with the International Organisation for Standardisation (ISO) and the Comité Européen de Normalisation (CEN), the European Standardisation Committee.

Selecting a course of action

Much will depend on the type of information needed and the country to which it is to be related. There are far more sources of information than can reasonably be listed, so selection must be on the application of logical thought based on an objective or plan. Each source of information selected should, with the right approach, lead to one or more further sources. Care is necessary to avoid unprofitable research so planning is essential and may involve considerable expenditure if the best information is to be obtained.

Chapter 14

Employment overseas. Selection, costs and service agreements

Basis of employment

Success with any international construction undertaking will depend on numerous factors and unknowns, but the selection of managers, technicians, craftsmen and manpower of all types is the most crucial. While home construction depends largely on skills, experience and management organization, overseas activities usually have the added complications of different climatic conditions, social customs and conditions, religious and a variety of factors separate from construction needs, but very much dominating influences. Staff selection, as a result, depends as much on assessing the abilities of those concerned, and their families, to adapt to these as on management and construction expertise. At all levels the need to be able to take decisions and for emotional maturity and stability can be as important as any other requirements. Key factors governing selection for overseas appointments are:

Expertise

Training and qualifications; professional, management or trade experience and overseas experience.

Suitability

Ability to work in country concerned; political and religious differences; family commitments and health, maturity and emotional background.

Implementation

Appointment and duration; contract of employment; travel; single and accompanied accommodation and domestic staff; food supplies and costs of living; insurance; medical; security; schools; social activities and transport.

Remuneration

Salary; allowances; taxation; method of payment; national insurance and pension provisions.

Conditions of employment

Many organizations will delegate recruitment and selection to personnel directors, managers and staff depending on the appointment. This may be from existing employees, and for senior management appointments there is much to be gained by using those familiar with the organization, policies and objectives of the undertaking concerned. It may not, however, be convenient or possible to do this, and external recruitment may be needed. This may be done directly or through recruitment organizations. The latter is particularly relevant when foreign employees are sought. Whatever method is used, the final selection is likely to involve assessment by the director or manager responsible and others dealing with specific functions.

Selection of designers to work overseas will depend partly on whether or not the appointment is for a specific office and, therefore, potentially long-term or for a project, which may be for a relatively short period. Those concerned with manufacturing are influenced in the same way, but may be recruited for lengthy contracts which involve virtual permanence overseas. For construction projects appointments are usually for relatively short periods, say for up to three years, but many international contractors have established a permanent presence in some countries and their recruitment of staff for such offices can be subject to similar influences to those governing manufacturers. Nevertheless, some appointments will inevitably be for very short periods, say for three or six months or a maximum of one year. Some craftsmen and technicians may wish to work on this basis only, as may sub-contractors or specialist contractors with relatively small contracts. Unskilled labour may be recruited locally on a direct basis or through agents. It may also be supplied from a third country, notably those with very high unemployment. Clearly wage rates can be a governing factor, but so, too, can religious and feeding factors. They may require the establishment of special services complicating further the already demanding tasks of mobilization and providing living arrangements.

Reasons for working overseas may be many and varied. For companies profit targets and home output are important. For individuals the reasons may be more involved and personal. Invariably finance and taxation are high on the list, but those concerned should realize that there are hazards which may involve a family. Employment overseas can also change a career structure in relation to work at home, and pension provisions and rights. The latter is less likely for short-term appointments or when a secondment is arranged, but should be discussed when selection is being considered.

For most contracts a work permit is likely to be needed. Remuneration may be by local and overseas payments, but a contract may also include housing, or an accommodation allowance, electricity and water, transport, and, for some, domestic staff. It may also cover messing for staff living on site on a bachelor basis, and usually payment of medical expenses. Travel costs at the beginning and end of the contract should be covered in a contract. Those for mid-term leave, local leave, in the event of sickness or family problems must be considered and may depend on the length of the tour and such factors as seniority. A move allowance may be paid. The contract, apart from protecting the employee, also protects the employer. Overseas construction activities demand far more than just job ability. Much may depend on the conduct of an individual when not working. This must be exemplary as any breach of local laws, misconduct or failure to observe accepted social requirements may jeopardize an existing undertaking and obtaining subsequent contracts or orders. A contract should, therefore, make provision for this, and should be a factor for discussion at selection meetings. There will be variations which make drafting a standard contract difficult. However, there are factors which are common to many activities. Specimen headings are:

- Appointment
- Location
- Period
- Reporting
- Agent for employer and powers
- Restrictions on other employments and gifts
- Salary, bonuses and allowances
- Travel costs
- Accommodation
- Domestic staff
- Medical services
- Personal transport
- Loss of remuneration due to employee's negligence or other causes
- Leave periods, qualification and exclusions
- Leave travel
- Termination of employment. Conditions and causes
- Travel arrangements in the event of termination of employment
- Repayment of allowances in the event of termination of employment
- Conditions affecting termination of employment due to ill-health
- Recoveries by employer in the event of termination of employment
- Special requirements. Qualifications and languages
- Law applicable

Taxation at home and overseas can be influenced by the way a contract of employment is drafted. Clearly this will depend on the appointment, location, duration and seniority of the person concerned. A suggested framework and one different from the headings already outlined is given below.

Framework for a contract for employment outside the UK

This agreement is made the .... day of .... 19 ....
Between .... whose registered office is situated at ....

Now it is hereby agreed that .... shall employ Mr ..... and that Mr .....
shall serve upon and subject to the following terms and conditions:

The employment shall be for a period of . . . . . years from the . . . . . day of . . . . . 19 . . . but subject to termination as hereinafter provided. Thereafter the employment shall continue from year to year until either party shall give to the other twelve (12) months notice in writing but subject in any event to termination as hereinafter provided.

Mr . . . . . shall exercise and perform such powers and duties as shall from time to time determine and in particular shall act as . . . . of . . . .

During the continuance of his employment hereunder Mr . . . . . shall, unless prevented by ill health, devote such of his time and attention to the business of . . . . outside the United Kingdom as the business of . . . . shall require having regard only to other contracts or engagements that Mr . . . . . may have with other subsidiaries or associated companies of . . . . and shall do all in his power to promote, develop and extend the business of . . . . and will at all times and in all respects conform to and comply with the directions and regulations made by the Board of Directors of . . . . and shall not engage in any other business or be concerned or interested in any other business of a similar nature to or competitive with that carried on by . . . .

Mr . . . . . shall (in addition to the usual public and legal holidays in . . . .) be entitled to . . . . . weeks vacation in each year to be taken at a time or times convenient to . . . .

Mr . . . . . shall not (except in the proper course of his duties hereunder) either during or after the period of his employment hereunder divulge to any person and shall use his best endeavours to prevent the publication or disclosure of any trade secrets or secret manufacturing process or any confidential information in turning the business or finances of . . . . or any companies or organisations to whom . . . . has provided services or any of the dealings, transactions or affairs of them and all notes and memoranda of such trade secrets or confidential information made or received by Mr . . . . . during the course of his employment hereunder shall be the property of the person originally making such available to Mr . . . . . and shall be surrendered by Mr . . . . . to someone duly authorised in their behalf at the termination of his employment or at the request of the Board of Directors of . . . . at any time during the course of his employment.

The employment of Mr . . . . . hereunder may be terminated by . . . . without payment in lieu of notice if Mr . . . . . is guilty of any gross default or gross misconduct in connection with or affecting the business of . . . . or in the event of any material breach or material non-observance by Mr . . . . . of any of the stipulations herein contained.

The remuneration of Mr . . . . . (which shall be deemed to accrue from day to day) shall be a fixed salary at the rate of . . . . per annum or such greater amount as shall be from time to time agreed payable by equal monthly instalments on the last day of every month and the first of such payments shall be made on the . . . . day of . . . . 19 . . . .

. . . . shall also pay Mr . . . . . all reasonable entertainment, hotel, travelling and other expenses wholly and exclusively incurred by him in or about the performance of his duties hereunder.

In the case of illness of Mr . . . . . or other cause incapacitating him from attending to his duties hereunder, Mr . . . . . shall continue to be paid during such absence, provided that if such absence shall aggregate in all twelve (12) weeks in any fifty two (52) consecutive weeks . . . . may terminate the employment of Mr . . . . . hereunder by notice given on a date not more than twenty eight (28) days after the end of the last of such twelve (12) weeks in which event . . . . shall pay to Mr . . . . . a sum equal to three (3) months salary from the date of such termination.

Nothing in this agreement shall require Mr . . . . . to undertake any duties or to work within the United Kingdom. It is hereby declared that the remuneration specified is solely attributable to work carried out and the discharge of duties and responsibilities outside the United Kingdom.

If before the expiration of this agreement the employment of Mr . . . . . hereunder shall be terminated by reason of liquidation of . . . . for the purpose of reconstruction or amalgamation and Mr . . . . . shall be offered employment with any concern or undertaking resulting from such reconstruction or amalgamation on terms and conditions not less favourable than the terms of this agreement then Mr . . . . . shall have no claim against . . . . in respect of the termination of his employment hereunder.

Mr . . . . . *Hereby covenants* with . . . . that he will not within one year after ceasing to be employed hereunder (without the previous consent of . . . . in writing under the hand of a Director duly authorised by a resolution of the Board) in connection with the carrying on of any business similar to that carried on by . . . . prior to such cesser on his own behalf of any person, firm or company directly or indirectly seek to procure orders from or do business with any person, firm or company who has at any time during the . . . . years immediately preceding such cesser done business with . . . . provided always that nothing in this clause contained shall prohibit the seeking or procuring of orders or the doing of business not related or similar to the business or businesses aforesaid or any of them.

If any clause or provision of this agreement is held to be invalid, such decision shall not affect the validity of any other clause or provision hereof.

This agreement shall be governed by the laws of England.

As *Witness* the hands of the parties hereof

Signed by
on behalf of . . . .
in the presence of:

Remuneration and personal taxation

Conditions for paying a salary, bonus, pension provisions, local allowances, education allowances, for travel and other provisions may be expected to be specified in a contract or service agreement. However, taxation, despite any advice which may be given by an employing organization, depends on the home taxation laws of an employee and on laws in the country of employment. Much will depend on the facts of employment and on interpretation of laws. The ultimate responsibility lies with the employee and inspector of taxes concerned. Taxation on payments received by a UK national will depend on the period employed overseas. The categories are for employment which involves short visits overseas of less than thirty days in an income tax year; employment overseas of more than thirty days in an income tax year, but less than 365 days; employment with an overseas employer when duties are performed abroad, irrespective of the period of the employment; employment covered by a contract of employment overseas for a period exceeding 365 days, but not covering an income tax year and employment involving absence for more than 365 days and covering a fiscal year.

Employment overseas for less than thirty days attracts no tax advantages, but for that exceeding thirty days there will be a deduction of 25 per cent of the foreign earnings when calculating UK tax liability. This may also apply when employment is undertaken overseas for a foreign employer irrespective of the period, but there are limitations. In the case of longer periods overseas, for 365 days but not covering a fiscal year, and for periods exceeding a fiscal year the definitions of domicile, residence and ordinarily resident in relation to the UK will be critical. Always it is essential to ensure that there is evidence of a contract for overseas employment and, when it is anticipated that residence status will be changed, it should be clear that the contract covers a complete fiscal year. It is often mistakenly assumed that an employee must remain out of the UK for a full fiscal year covered by a contract. This is wrong as it is the contract of employment which must cover the fiscal year. During that period an individual may return to the UK for short leave periods. When there are doubts the Board of Inland Revenue should be consulted, or the tax inspector concerned, as they rely heavily on a code of practice in the absence of statutory guidance.

Employment overseas for periods exceeding 365 days but not covering a fiscal year

Provided overseas employment is for a continuous period which exceeds 365 days there will be no UK income tax liability on the emoluments from the employment. The duties must be performed overseas, although there is an exception for duties which are carried out in the UK which relate to the employment. Periods which the individual concerned can spend on visits to the UK must not be for more than sixty two consecutive days and must not exceed one sixth of the qualifying period of absence. Effects on UK tax liability depend on the number of days spent outside the UK in foreign employment. The fiscal year is not relevant to the determination of the relief. An individual with a contract who satisfies these conditions continues to be resident and ordinarily resident in the UK for taxation purposes and should continue to make annual returns of income, but taxation liability on the actual overseas earnings should be relieved in full. Income from other sources is not affected in the same way.

Employment overseas for a period in excess of a fiscal year

Evidence of a contract for the overseas employment is necessary. This may take the form of a letter from an employer stating that the contract will be performed overseas. Taxation may vary if some of the duties are performed in the UK. There should also be evidence of the period covered by the contract, and it should be clearly established that the period of the contract covers a full fiscal year. The inspector of taxes in the UK to whom the individual makes income tax returns should be advised of the overseas contract prior to departure to establish the change of residence status. During the period of the overseas employment the individual should not visit the UK for periods exceeding three months in any income tax year in order to qualify for full tax remission.

Having established that the contract satisfies these conditions the individual concerned will cease to be resident and ordinarily resident for taxation purposes in the UK from the date of departure and will be non-resident for the period of the overseas employment. There will be no tax liability in the UK on the foreign earnings during the period of non-residence, or on other foreign income nor liablility for capital gains tax on profits from disposals of assets during the period of non-residence. There will, however, be liability to UK tax on any income arising in the UK, but there could be relief under a double taxation treaty between the UK and the country of overseas employment. It must be noted, however, that the taxation status of the wife of an individual employee in these circumstances usually remains unchanged even though she may take part in the overseas tour of duty.

Individuals working overseas and returning for temporary employment to the UK

When individuals working abroad under a long contract return temporarily to the UK to carry out duties for a comparatively short period the taxation consequences need careful study. The employee would normally cease to be resident and ordinarily resident on taking-up overseas employment. Whether or not a period of temporary employment in the UK affects taxation depends upon the period of time, but in all such cases there would be UK tax liability on remuneration for services performed in the UK. The residential status is important and depends on whether an individual returns to the UK for temporary employment for a period of less than two years, for between two and three years or in excess of three years.

Final selection

Time spent selecting the right staff for overseas appointments can be lengthy and frustrating. Good selection will, nevertheless, govern the success of an undertaking. Those experienced in overseas activities will already know the problems of working, living, travelling overseas and their impact on family life, but those taking an appointment overseas for the first time have no basis for comparison. They must understand what is involved and their reactions must be tested at an interview with those responsible, possibly at several interviews. Failure to adapt to overseas needs or to family pressures leading to a broken contract can only be disruptive to all and certainly expensive.

Chapter 15

Procurement, invoicing and documentation for overseas contracts

Buying policies

Products to be used in a construction project will normally have been specified by the designer, and precise descriptions of such products will have been given in the bill of quantities from which the contractor prepared the relevant bid. Equipment, furniture, machinery, tools and other products may also have been specified, and will be an important part of any buying programme when this is for a turnkey project. Even when a contractor is concerned only with a structure the end-use can influence working and output targets. Some products will have been specified so precisely that purchase can only be made from a nominated supplier to a given specification. Others will permit selection by the designer or contractor, but always subject to meeting specification, quality, performance and price. The role of the project director or manager will be as much dominated by this as any other matter, but a project procurement manager or a centralized buying department is likely to have the most direct and effective role.

A buyer is always likely to face pressures to buy at a better price, more quickly and to meet delivery dates that more often than not have been overlooked by some other person. Too often the design will have been prepared many months or years earlier, possibly by a consultant unfamiliar with the country in which the project is to take place and certainly with no knowledge of the eventual contractor. It may also incorporate products which are out of production, obsolescent and frequently higher in price. Even with provisions for price-escalation, as a result of specific percentage related to inflation in either the country of purchase and origin stated in a contract or indexing in some other way as a result of inflation or price increases generally, there will be constant pressure to keep to a bid price and profit target.

Competition for international construction contracts is such that some contractors may bid for work at unrealistic prices, possibly encouraged by some governments who support a contractor working overseas for political reasons or to encourage employment, products sales or as a means of obtaining foreign earnings. In these cases a buyer will face even greater problems. So, too, will nominated suppliers who may be pressured to reduce prices. This is particularly likely if the contractor has been able to obtain the agreement of the consultant for alternative products to those specified. There are some contractors that under-bid, and are awarded contracts, on the assumption that they will be able to profit by subsequent worldwide keen buying, and by changing specified products to those with lower prices.

Programme

Procurement will have been considered in detail when bidding for a project, and with some anticipation of price increases. Usually, however, time allowed for bids fails to permit the comparative pricing needed. Much will depend on the speed of reply by suppliers and price information held by a contractor. In many cases computers and retrieval systems will be used, but information from these is always likely to be out of date. On award of a contract some products will be needed immediately, such as those for mobilization and for work below ground. Others must be phased to a work schedule. As an example some services products will be needed at an early stage and others towards the end of a project. Finishing products and some equipment may not be ordered until a very late stage and will not be installed until a short time before hand-over so that damage is prevented.

Whatever the programme it is unlikely that bid prices will be relevant by the time a project starts. The buyer must, therefore, obtain revised prices and new bids. In each case a price and purchase will be influenced by delivery time, validity, method of payment and the buying and selling conditions involved. Some will order and buy products in advance of needs hoping that discounts and rising prices will offset storage and financing costs. This is less likely than phased purchasing, usually later than needed with consequent needs to improve delivery. Even when delivery has been stated in a quotation by a manufacturer as, say, ten weeks, and a price validity of ninety days given perhaps six months earlier, some contractors working in international markets will expect prices to be held and delivery to be less. Market conditions may force a supplier to meet price pressures, but delivery times depend on many production planning factors.

Payment for products bought on a worldwide basis may be by a cash transfer by mail or telegraph, but letters of credit will frequently be used. These give protection to a buyer and to the seller. Delivery will depend on the freight arrangements made by the supplier, buyer and forwarder. These, too, must be planned and priced in advance. Late delivery or a changed buying programme can disrupt a freight price and can, in turn, change a delivery significantly. When the relevant construction project is remote and in a developing country which has distribution problems and limited shipping, air or road services changes can be critical and lead to expensive charter arrangements. Increasingly buying is more scientific and based on sophisticated calculations and information storage, but these and purchase are still governed by laws of supply and demand and negotiating skills. The latter, particularly, by problems arising and such imponderables as scarcity value. Effective buying should overcome as many of these factors as possible, and depend on sound planning.

Method of working

An actual buying plan may be controlled directly from a construction site overseas, through a home office or a combination of both. Some contractors, particularly those from developing countries, may use confirming houses, or merchants and procurement agencies from their own country or from elsewhere. This may be particularly necessary when precise specification has been made, such as use of products to a British Standard (BS) and when specialized technical knowledge and an understanding of specific manufacturers is needed. However, most UK contractors will normally use their own buying resources and only resort to external agencies when a large buying programme is involved or in the initial stages of operation in a new territory. In all cases buying will depend on trading or commercial skills, but these must ideally be supported by specialized training and knowledge of construction. A procurement or commercial education is one part of the needs, so, too, is technical training, for example when electrical, mechanical and scientific products are being bought. Architectural training can help with purchase of many products. Those concerned should ideally understand standards, such as BS, and Agrément Certificates, both of which are important to buyer and seller. Quality control certificates may also be relevant. However, in cases such as the bulk purchase of cement, steel, pipes and similar products freight, pricing and commercial skills may be more necessary than technical knowledge, provided a proper specification is available and is understood. All trade between the buyer, supplier, carrier and forwarder and payment depends on effective export invoicing and documentation.

Invoicing

Suppliers may be expected to quote and invoice in their own currency, having calculated all costs and a profit margin. Some exporters, like those from the UK, may invoice in another currency. They should, before doing this, ensure that the currency will not be blocked or made inconvertible in any way. There may be administrative reasons why invoicing on one's own currency is convenient, but there may be exchange fluctuation risks with this or any currency selected. For those with limited cash flow or ability to face exchange risks, factoring or selling an invoice at a discount may be necessary. In any event the highly sophisticated London forward exchange market will enable a seller to assess and anticipate currency fluctuations and, therefore, a potential gain or loss. Some buyers may insist on paying in a specific currency to suit their trading requirements or assessment of exchange rates. This could be to the advantage of an exporter. If the exporter invoices a buyer in a foreign currency, for payment at a future date the sale proceeds in that currency can be sold forward to a bank through the forward exchange market. The forward rate for the currency, in sterling, may be at a premium or at a discount against the spot rate of exchange depending on market. Forward cover may needed to protect against default.

Documentation

Careful export documentation is always needed, but never more so than in the case of documentary credits. Uniform Customs and Practice for Documentary Credits published by the International Chamber of Commerce is an accepted code of practice. A variety of documents are needed to support the origin of goods, tariff preferences, proof of shipping, title and eligibility for payment apart from a commercial invoice from the supplier to the buyer. An incorrect document can lead to complications and delayed payments. Within the EEC there is standardization of documents influencing the free movement of goods which is known as the Community Transit system (CT), but for most international construction markets the export documents needed follow a well established pattern and are defined below.

Bill of exchange

An unconditional order in writing, addressed by one person to another, signed by the person giving it, requiring the person to whom it is addressed to pay on demand or at a fixed or determinable future time a certain sum of money to or to the order of a specified person, or to bearer. The exporter may draw a bill of exchange on a buyer and pass it with shipping documents and collection instructions to his bank which forwards the bill and documents to a bank in the buyer's country. This would present a sight bill of exchange to the buyer for immediate payment or a term bill for acceptance. Should the buyer refuse, the documents would be withheld and the control of the goods remain with the bank and seller. When a letter of credit has been opened, the bill of exchange must be in accordance with the terms of the credit and will normally be drawn on the bank through which the credit is opened, advised or confirmed.

Bill of lading

A bill of lading is a receipt for goods and evidence of a contract of carriage. It is also a document of title enabling an exporter to transfer ownership or possession of goods, usually in sets of two or three originals any one of which gives possession of the goods. Among the particulars are the name of the shipper, the date and place of shipment, the name of the vessel, the port of destination, a description of the goods and the shipping marks. It usually states whether or not freight has been paid and may bear the name of the consignee or that the goods are shipped to order. A shipped or on-board bill acknowledges that the goods have been loaded and a received-for shipment bill confirms that the shipping company or its agents have the goods in custody for shipment. The latter is not normally acceptable under the terms of a letter of credit, but it may be converted to a shipped bill by the carrier after the goods have been loaded. A clean bill is one which has no clause or statement declaring a defective condition of the goods nor of the packaging. A through bill may be issued when different modes of transport are used and when there is no direct shipping link which necessitates arrangements for the goods to be transferred to a second ship at another port. A container bill of lading may also be used in the case of container shipment.

Certificate of origin
Evidence for the importing country that the goods originate from a country from which imports are permitted.

Combined transport document
Covers movement of goods by two or more different methods of transport.

Commercial invoice
Describes the goods, states price and buyer. Other details are quantities, weights and measurements, packing details, shipping marks and terms of sale. These must conform to the contract of sale and documentary letter of credit if opened. Several copies of the invoice may be required. A combined invoice and certificate of value and origin may be used for shipments to some common-wealth countries.

Consular invoice
Invoice issued in the country of an exporter by the consulate of the importing country to support freight details for the authorities in the importing country.

Consignment note, waybill and air waybill
These are provided by a carrier when goods are sent by air, rail or road and are receipts for the goods and evidence of dispatch to a consignee. The extent to which these documents may be used to effect the transfer of ownership or possession of goods varies.

Insurance
An insurance certificate may be required to support documents.

Contractual requirements

The main contract, that with a sub-contractor and matters like insurance will influence procurement as the contractor is responsible for the fitness of a product for use, its installation and during maintenance period. When a contractor has given a performance bond to an employer he may in turn require a supplier to give a performance bond, which is usually resisted. However, when a large supply contract phased over months or longer is involved a manufacturer is likely to accept the conditions imposed, particularly when an advance payment is made. A supplier may insist on an advance payment when a large order is involved. This also applies in the case of capital equipment designed and manufactured for a project.

In the absence of contrary agreed conditions a supplier will normally state that a quotation or supply is based on conditions which exclude those of any other. It may be necessary to assess these in relation to those of a carrier.

Example terms and conditions of sale. Manufacturer

General

Orders are accepted subject to the following terms and conditions. No variation or modification of, or substitution for, such terms and conditions shall be binding unless expressly accepted by the Company in writing.

Quotations

Quotations do not constitute an offer by the Company to supply the goods or carry out the work referred to therein, and no order placed in response to a quotation will be binding unless accepted by the Company in writing. All acceptances by the Company are subject to availability of materials and to the Company being able to obtain any necessary licences and to the same remaining valid.

Prices and terms of payment

All prices given in quotations are provisional until the order has been accepted by the Company in writing and being based upon the prices and costs of materials, labour, transport and other expenses at the date of quotation may be varied by the Company before or after acceptance of the order to correspond with any variation in such prices or costs which may occur at any time before delivery of the order to the customer. Unless otherwise stated prices quoted are net. In addition to any right of lien to which they may be by law entitled, the Company shall in the event of the customer's insolvency be entitled to a general lien on all goods of the customer in the Company's possession for the unpaid price of any other goods sold and delivered to the customer by the Company under the same or any other contract.

Exclusion of warranties

While every care is taken in providing suitable goods and in giving particulars of capacity and performance, the Company gives no condition or warranty as to their quality or fitness for any purpose whether that purpose is known to the Company or not and all express or implied warranties or conditions statutory or otherwise are expressly excluded. The Company shall not be liable for any loss or consequential damage.

When the Company gives advice or approval concerning plans or specifications or concerning any other matter in relation thereto such advice or approval is given subject to the condition that the Company shall be under no liability of any kind.

Drawings, descriptions, weights or dimensions submitted by the Company are approximate only and are intended as a general guide. The Company will not be liable for any error or omission therein or with regard thereto.

In the event of any defect in any goods supplied by the Company due to faulty material or workmanship the Company will, if satisfied that such defect is due to faulty material or workmanship, repair or replace such goods provided that notice of such defect is given in writing to the Company within fourteen days of delivery and the goods are returned carriage paid to the Company's works within twenty-eight days thereafter. After the expiration of such period all further liability on the part of the Company shall cease. The Company's obligation to repair or replace such goods is subject also to the customer having complied with all instructions given by the Company concerning the manner in which such goods should be used.

Patents and designs

The Company shall not be liable in respect of any claim which may be made against the Company for infringement of letters patent or registered design arising as a result of the Company carrying out instructions given by the customer and the customer agrees to indemnify the Company from and against all or any such claims and against all costs, damages and expenses incurred by or recovered against the Company in respect of such claims.

Delivery

Unless otherwise agreed delivery will be made at the Company's works. Delivery dates given by the Company are approximate only and no liability can be accepted for any loss, injury, damages or expenses consequent upon any delay in delivery, nor shall any such delay entitle the customer to cancel any order or to refuse to accept delivery at any time.

Damage or loss in transit

The Company will not be liable for loss or damage to goods or materials in transit unless written notice is given to the Company in accordance with the terms and conditions of the insurance cover effected in respect of the particular consignment. Notice must be given in the case of loss within seven days of the scheduled arrival date or in such other manner as may be specified in the covering documents.

Packing

Except when the Company exports out of the United Kingdom, the customer shall return carriage paid to the Company's works in good condition within one month of receipt all packing cases, drums and crates provided and invoiced as returnable. The customer shall not make any deduction in respect of the cost of such cases, drums or crates, but credit will be allowed up to the full amount charged therefore if returned and accepted by the Company within the time specified.

Resale

If the customer shall sell any of the goods purchased from the Company he must do so under the trade marks or trade names registered by the Company, and the customer must ensure that goods are only sold subject to the terms and conditions as are herein contained unless otherwise expressly agreed in writing by the Company.

Force majeure

Should the Company be delayed in or prevented from making delivery owing to act of God, War, Civil disturbance, requisitioning, Government or Parliamentary restriction, prohibition or enactment of any kind, import and export restrictions, strikes, lockouts, trade dispute, difficulty in obtaining workmen, or materials, breakdown of machinery, fire, accident, or any other cause whatsoever beyond the Company's control, the Company shall be at liberty to cancel or suspend the contract without incurring any liability for any loss or damage resulting therefrom.

Interpretation

Quotations and contracts arising therefrom shall be construed according to and governed by English law and the customer hereby agrees to submit to the exclusive jurisdiction of the English Courts in any dispute or difference of any kind which may arise concerning any quotation or contract made between the Company and the customer.

Extent of conditions

The foregoing terms and conditions supersede and exclude all general or special terms or conditions imposed or sought to be imposed by the customer at any time in relation to any order arising out of this quotation.

Licensing agreements

Equipment, manufacturing machinery and products may be purchased through a licensing agreement. This may be for a patented product, but often will be a way of paying for the transfer of skills which are the result of investment and experience not subject to patent, such as the manufacture of a complete building system and assembly techniques. This is far more intangible than an agreement for the use of a specifically designed item of machinery or a product which has unique characteristics.

A licensing agreement may provide for outright purchase, a basic payment and royalties or royalties only. It could be for an indefinite period, but will normally be for a specified period and restricted to a particular country or region, although it could be worldwide. An agency may be permitted on a sole or exclusive basis, but this is not permitted in countries of the EEC. The agreement will almost certainly provide for a number of services by the seller covering such functions as: installation of equipment, training of operatives and subsequent supervision for a stated period, recruitment of managers and operatives, local market research and selling advice, manufacturing materials purchasing and control, modifications and possibly design and management of a factory in which the machinery is to be installed. Much will depend on how an agreement is drafted and in the case of royalty payments the basis for supervising output figures. This will often depend on goodwill. Some form of inspection is necessary, but is not always possible or, indeed, permitted in some countries.

Competition governs price

There are few products used in construction which are not subject to intense worldwide competition. Advanced products such as some machinery and capital goods like electrical equipment which may be of very high value can face less competition than, say, architectural products, although financing costs, long-term maintenance and national prestige exert different pressures. There may also be strategic limitations. In developing countries where there are large markets for construction many products may be imported with little or no tariffs, but as such countries develop and start manufacturing their own construction products insistence on using locally made products may result and tariffs may be created or raised. The price and quality of such products will have an important influence on buying and construction, and the trend is one which will have a growing impact on future markets and the export of products from the UK and other industrial countries.

Chapter 16

Payments, financing methods and documentary credits

Cash flow

Irrespective of by whom, how and where payment is approved the actual method of payment requires careful planning and supervision, and will dominate the cash flow of any undertaking and, therefore, its ultimate profitability. It will influence cash in hand, its use and investment and may create a need for short- or medium-term loans or conventional overdraft facilities. In some cases advances against foreign bills may be negotiated and term drafts permitted under a documentary credit may be discounted. Factoring, hire purchase, leasing and other methods may be convenient or even necessary for some. Cash flow may also be supported by credit given by suppliers, merchants and confirming houses. Insurance cover for financing arrangements is an important part of the services of the Export Credits Guarantee Department (ECGD) which may provide a comprehensive bank guarantee, pre-shipment finance guarantees or cover for medium-term buyer credit to enable bank financing to be arranged. Provision of finance is usually, however, a matter for bank negotiation by all concerned, and will largely depend on the creditworthiness of those involved and the project apart from insurance which is available.

Any construction activity must depend on the ability of a buyer or employer to pay, and, assuming that the source of finance, buyer or employer, project and factors within the country concerned are acceptable, a designer will need a steady payment of fees, a contractor must ensure payments against certificates and suppliers must be confident about payment before manufacture or sale. A letter of intent and acceptance of an invoice may give some confidence, nevertheless cash transfers or documentary credits are likely to be the only methods generally acceptable.

Terms and methods of payment

Regulations may influence exports from the UK. These are known to banks, and are available from HM Customs and Excise offices. Strategic limitations may apply and could affect trade with some countries. More significant, however, are the ways in which local import regulations will influence the ability of a buyer to pay. For these and other reasons establishing the terms of delivery or project completion dates are crucial, particularly for manufacturers who will also influence the performance of a contractor. Delivery of goods free on board (FOB) is often the easiest method, but a foreign buyer may insist on a contract cost and freight (C & F), cost, insurance and freight (CIF) or to a stated location. Quotations must allow for all costs additional to FOB costs. Depending on the size of the contract and trading method, payment may be in advance of shipment, when goods are shipped, at an agreed time after shipment or after delivery and acceptance of the goods. There may also be arrangements for a periodic settlement of outstanding invoice values and combinations of methods of payment, for example 20 per cent of a total contract value may be payable in advance, further 60 per cent of the value of each shipment at the time of shipment. and a final 20 per cent of the full contract value after acceptance of the goods or at an agreed date after final shipment.

Payment may involve a transfer of funds by a buyer at an agreed stage of a transaction, possibly by telegraphic transfer through a bank or by the opening of a documentary credit by a buyer in favour of a seller, probably confirmed by a UK bank. On presentation of shipping documents in accordance with the terms of the credit it is possible to obtain immediate payment under a sight credit or a bank acceptance in the case of a usance credit, or by a buyer paying a sight draft or accepting a usance draft drawn on the buyer by the seller. Payment may also be by negotiating the relative draft and export documents with a UK banker, by selling export invoices to an export factor or by securing an advance from a banker against a guarantee issued on behalf of a seller by ECGD, who may also have provided a buyer credit facility. A UK bank may also provide a foreign currency loan against eventual export receipts.

Invoices may be in sterling which has administrative advantages for UK sellers, but a foreign buyer may prefer to be invoiced in another currency. Foreign currency invoicing could involve an exchange risk, or a bonus depending on the strength of sterling on receipt of payment and exchange. To minimize potential exchange losses it is possible to sell anticipated currency receipts to a UK bank forward, by which a bank will fix in advance a rate of exchange to be applied when the currency is received. This arrangement is a contract. If the anticipated proceeds are not received by the due date it will be necessary to purchase the currency at the spot rate to meet the obligation to the bank. Other risks are when a buyer defaults or becomes insolvent, when an importing country prevents or delays payment, frustration of a contract due to war or political disturbance, new import licensing or exchange control restrictions imposed in a buyer's country and other causes of loss unless they are insurable which are outside the control of the buyer or seller. These may be covered by export credit insurance with ECGD or a private sector credit insurance company.

Documentary credits

Although relatively simple in concept, payment through use of a documentary credit depends on conditions written into the credit, supporting documents, correct timing and efficient banking methods. The probability of error is, therefore, considerable and, as the increasing use of documentary credits owes much to growing and complex world trade, this is compounded by the relative efficiency of all involved. The basis of a documentary credit is to enable a bank acting for a customer to pay to a third party, known as a beneficiary, a specified sum, usually through a bank, in a country or place nominated by the beneficiary. This bank is known as the correspondent bank. It may also be, in addition to being the paying agent, the advising and confirming bank. The process is known as opening a letter of credit. There are thus four parties involved: the buyer or applicant for the credit, the issuing or buyer's bank, the paying agent and advising or confirming bank and the seller or beneficiary. Despite the minimizing of risks by using a documentary credit the creditworthiness of a seller should still be investigated prior to agreeing a contract.

There are several types documentary credit, there is also the traveller's letter of credit and the cash letter of credit which are more likely to be used for personal transactions, which give varying degrees of protection. They are summarized below. In the Far East an authority to purchase may be used. This is not strictly a letter of credit, but functions in a similar way.

Types of credit

Revocable credit

Permits amendment and cancellation of goods, even when in transit, before presentation of documents and before payment is made after presentation of documents. It gives a buyer maximum flexibility, but little protection for the seller or beneficiary.

Irrevocable credit

This cannot be revoked, but payment can be dependent on the opening bank which is likely to be foreign. The buyer has limited flexibility as the credit can only be amended or cancelled if the parties involved agree.

Confirmed irrevocable credit

In this case a bank in the country nominated by the seller for payment adds its undertaking to the of the issuing bank. This bank is likely to be in the seller's own country thereby giving considerable protection. Using this type of credit involves more expense than the other.

Payment documents

A credit issued may call for the presentation of sight drafts in addition to specified documents. The drafts are likely to be required to be drawn on the buyer, issuing bank or paying bank. If the seller agrees to selected payment facilities by the buyer a usance draft could be used requiring the beneficiary to present documents at a specified period after shipment. In this case payment is agreed for a given date and permits acceptance by the negotiating bank on behalf of the issuing bank. A deferred payment may be stipulated in conditions of a documentary credit. They may also provide for the delivery of goods and payments to a given programme, which would be known as a revolving documentary credit. In some cases advance payments may be permitted. This is particularly important when the credit is for expensive capital goods or when a large consignment of one type of product or a collection of products is assembled for one shipment.

Transfers, back-to-back credit and assignment of proceeds

When a seller and beneficiary of a documentary credit is not the manufacturer, say in the case of a merchant or procurement agency, payment may be needed before release of goods. This may not be convenient unless cash or credit facilities are available and it may be necessary to transfer funds from the credit. This can only be done if it is transferable credit. The most secure and flexible documentary credit is, therefore, the confirmed, transferable irrevocable letter of credit. It can be made more flexible if conditions drafted allow, such as by permitting partial shipment, transhipment, and transfer of funds in respect of FOB value and freight value in selected cases. If transfer is not provided for, a beneficiary may ask the advising or confirming bank or other bank to issue a new credit in favour of the supplier. This is known as a back-to-back credit, and need not be known to the final buyer. Alternatively it may be possible to arrange an assignment of proceeds of a documentary credit. In both cases the banks involved must be assured of payment under the first credit. Examples of instructions for the total and partial transfer of funds are as follows.

Application for the total transfer of funds

.... Bank   Date ....

.... Branch   Advice number ....

Documentary credit reference ....

For the amount of ....

Drawn in favour of ....

Opening Bank ....

For the account of ....

We hereby irrevocably transfer all of our rights under documentary credit number ....

to ....

....

The transferee (the second beneficiary) shall have sole rights as beneficiary under the credit including all rights relating to any amendment whether increases or extensions or other amendments and whether now existing or hereafter made. The credit hereafter may be amended, extended or increased without

125

our consent or notice to us and you will give notice thereof directly to the second beneficiary. The original credit is returned for your endorsement.

We ask you to notify the second beneficiary of the transfer of this credit and of the terms and conditions of the credit. We agree to indemnify you against any and all losses damages and expenses arising from this transfer.

Authorised signature
Name of first beneficiary

Accepted by
Authorised signature
Name of second beneficiary

Application for the partial transfer of funds

.... Bank   Date ....
.... Branch   Advice number ....

Documentary credit reference ....
For the amount of ....
Drawn in favour of ....
Opening Bank ....
For the account of ....
We hereby irrevocably transfer in part documentary credit number
.... to ....

....

under the same terms and conditions of the credit except that the amount to be transferred is ...., the latest shipping date is ...., the expiry date is ....and the quantities are ....
Our special instructions are ....

Any amendment except an increase of the amount of the credit hereafter made is to be advised to the transferee (the second beneficiary) with the same notice to us. The original credit is attached. We ask you to endorse this transfer and return it to us. We agree to indemnify you against any and all losses, damages and expenses arising from this transfer.

Authorised signature
Name of first beneficiary

Accepted by
Authorised signature
Name of second beneficiary

Charges and commissions

All aspects of opening and handling a documentary credit, including transfers of funds, involve charges and the payment of bank commissions. There is likely to be a quarterly charge for opening an irrevocable credit and for a confirmed, a notifying commission and an issuing commission. There will also be a transmission charge. On negotiation, that is to say on presentation of the relevant documents for payment, there will be a commission charged for the examination by the bank concerned and charges for any subsequent postal and similar expenses. Any deferred payment will also give rise to a commission. The involvement of a second bank will incur their charges and commissions. Charges will also arise when there is a transfer of funds.

Supporting documents and conditions

Documents needed to support payment under a documentary credit are transport documents, such as a bill of lading, combined transport document or air waybill; consignment note; certificate of receipt; commercial invoice; certificate of origin; certificate of inspection and insurance certificate or combinations of these documents. In some cases a forwarder's certificate may be needed to support other documents. Some documents may require endorsement by an embassy or consulate in the seller's country and by a chamber of commerce. They are invariably needed as a part of any export transaction and as a part of a procurement programme, and must be carefully prepared and checked in relation to the requirements of a documentary credit to ensure that they meet stipulated conditions such as dispatch date, quantities, price, endorsements, and presentation and also conditions such as those for partial dispatch, transhipment, freight costs and payment and any transfer of credit. The negotiating bank must be assured of the fulfilment of all conditions before payment is made. If at any time a seller thinks that dispatch or negotiating dates require change the buyer and bank concerned should be told. This is particularly important in the case of an irrevocable documentary credit, but it must be remembered that changes involve expenses. A seller as the first beneficiary must, therefore, check the conditions of a credit on receipt to ensure that they meet those agreed as a basis for sale and that unforeseen factors have not made them impracticable.

Rules for using documentary credits

The need for legal protection or, at least, codes of practice to protect the use of documentary credits is widely recognized by many countries. Their value is so important to world trade that the International Chamber of Commerce (ICC) has prepared rules relating to current practice in an attempt to give a framework for their use. These are known as Uniform Customs and Practice for Documentary Credits. The matters covered are: general provisions and definitions; form and notification of credits; liabilities and responsibilities; documents generally and documents evidencing shipment or dispatch or taking in charge; marine bills of lading; combined transport documents; other shipping documents; insurance; commercial invoices and other commercial documents; miscellaneous provisions such as for quantity and amount; partial shipments; shipment, loading or dispatch; presentation; date terms and transfer. The ICC has also prepared standard forms for issuing documentary credits covering requirements.

127

Creditworthiness

No payments procedures can ever give perfect protection, and much will depend on how each organization and individual in the sequence between negotiation of a purchase, whether for expertise, goods or a service, meets the relevant part of an agreement or contract. A buyer will need to know that the seller is able to meet agreed obligations and the seller must be assured of the ability of the buyer to pay. In between are those who provide supporting services and who, in turn, need to ensure that they will be recompensed for them. Assessing efficiency and creditworthiness is, therefore, an essential factor in payments arrangements.

Chapter 17

Lending agencies and regional trading

Multilateral aid

Construction, particularly international projects, is frequently dependent on government or consortia financing. Even private sector projects are rarely financed by one source of funds. As the size of projects increases and those involving advanced technology, whether for manufacturing or for the development of infrastructures, lead to higher and higher capital costs the need for consortia or national financing increases. Often this is beyond the facilities available to the country concerned and international financing may be necessary. Projects in developing countries are typical of this need, but some in already highly industrialized countries may be financed on an international basis. Tunnels through the European Alps are examples, so, too, could be a Channel tunnel or bridge linking the UK and France which would benefit Europe as a whole and, indeed, countries further field. An understanding of international lending agencies and multilateral aid generally on which so much of international construction can depend is an essential part of any approach to markets.

Bilateral aid

The growth of multilateral aid is comparatively recent, largely during the past thirty years, but bilateral aid in various forms has a long history. It can be related to political and diplomatic objectives as much as to trade and genuinely helpful motives. One of the major providers of bilateral aid is the USA, but European countries also make a great contribution to the development of other countries. In many cases their aid is a result of former colonial or dependent territories links, notably in the case of France, Spain and the UK. The UK gives widespread support for projects in former colonial territories and to existing members of the Commonwealth, among others. Many UK government departments can be involved, but the Overseas Development Administration (ODA) has a major role, and can be consulted by those interested in particular markets. The ODA publishes details of projects on a regular basis. Equally important are their relationships with the Foreign and Commonwealth Office and the Department of Trade, and their projections for future development aid.

World Bank

One of the most important organizations concerned with multilateral aid is the World Bank. There are three international bodies within the World Bank group; the International Bank for Reconstruction and Development (the World Bank) and two affiliates, the International Development Association (IDA) and the International Finance Corporation (IFC). The UK is a member of each of these. The role of the group is to provide funds or to guarantee loans made by others to aid economic development in member states. There are about 130 countries now members of the World Bank whose original Articles of Agreement were drawn-up at the Bretton Woods conference of July 1944. The guidelines for operations give priority to urgently needed productive projects, but not to lend money in cases where funds could be obtained from other sources on reasonable terms; to considering the prospects for repayment before making a loan and ensuring that borrowers use funds for the purpose for which a loan is made. The bank still adheres to these principles, and was one of the first of a new type of international investment bodies. It uses its own capital, raised by contributions from members to the share capital of the bank, retained capital and that from bond issues. Subscriptions by members depend on the financial strength of the applicant, which restricts their share of a risk proportionately in the same way as any other investment. In this case, however, they also stand to gain by loans for development in their own countries. The World Bank is managed by a Board of Governors, one from each member country, through an international staff. Their role and operations evaluation in selecting projects for financing and subsequently has a strong influence on international construction markets.

Cash for projects may also be raised through World Bank support for co-financing. This is an arrangement by which World Bank funds are allocated with those raised from other sources outside the borrowing country. They may come from governments or their agencies, other international institutions, export credit bodies or private sector financial institutions. In the case of the latter joint stock and merchant banks, insurance companies, pension funds and similar sources may participate. United Kingdom experience in arranging such funding is among the best available and is widely respected by governments and industries throughout the world. Co-financing may be necessary when the initial contribution by the World Bank and borrowing country is not sufficient and when additional expenditure arises on a project, such as the need for operating supplies, to meet other initial costs and for working capital.

Project selection

The range of projects can be considerable, from construction needed to improve social conditions to advanced infrastructures or industrial and other developments such as agriculture, electrification, forestry, hotels, housing, hospitals, natural gas pipelines, ports, railways, roads, telecommunications, urban development and water supply. Usually, a project must be in a member country or one administered by the member, and the bank must be satisfied that it will be properly managed before and after completion, that a loan can be repaid and interest payments met and that the borrower cannot obtain finance at a reasonable rate elsewhere. Clearly finance for similar developments in different countries will be viewed differently depending on the stages of development and finance available. Loans cannot be given for armaments expenditure. Much will depend on how the World Bank assesses the prospects for an applicant country, but the bank may also suggest projects to a member as a result of staff assessments or those by such bodies as UN agencies. The factors and principles considered by bank specialists after identification of a project are very similar to those used by others and are listed below.

Economic
Benefits from project including and services, employment and contribution to the gross domestic product, gross national product and balance of payments.

Technical
Feasibility studies, design, location, integration with other developments, costs and construction, management and staff available.

Management
Staff available for managing project after completion and training needed prior to, during and afterwards.

Commercial
Resources needed and for subsequent use. Impact on suppliers and purchasing costs locally and worldwide.

Financial
Funds available, needed and subsequent running costs and income. Impact on other projects.

Design and tendering

Following approval in principle of a project a designer may be commissioned by the World Bank to undertake a feasibility study, a detailed project or implement a project. The normal procedure for consultancies financed by World Bank loans is for the borrowing country to draw-up a short-list of consultants and submit them to the bank for approval. It is essential, therefore, that consultancy firms wishing to participate in World Bank projects should register with the bank. The stage after design is for a project to be advertised for international competitive bidding, which is open to those in member states and Switzerland. This may be done by the borrower, who is responsible for the project, or by

131

another agency carrying out the project. However, the World Bank will wish to see that arrangements are effective and does not permit a borrower to prevent pre-qualification, although this will clearly act as a means of pre-selection prior to bidding. Contracts offered may be for construction or supply, and the bid documents will state whether or not the contract is for a lump sum, a unit price or a combination of these. When possible separate contracts are awarded for the civil works and major items of equipment. Turnkey contracts are usually only awarded when a closely integrated manufacturing process is involved. In the case of a turnkey project detailed technical specifications are published before an invitation to bid, and unpriced technical bids are usually required before priced bids are submitted.

Normally only part of the cost of a project is financed by the bank, and the allocation of that part of a contract financed by the borrower must be agreed between the borrower and the bank. It will not, however, agree to compulsory joint ventures by external organizations with local firms, but may agree a local or regional preference for products and a local preference for contractors when these are viable and when such decisions would further encourage development. An alternative to international competitive bidding may also be agreed if the value of a contract is too low to interest foreign bidders, when there are limited suppliers of particular goods and when there is a case for standardization. The types of factors which are likely to be detailed in an invitation to bid and which must be considered when bidding for a World Bank project are outlined below. While these follow a familiar pattern, it is as much on the objectives, principles of working and financing methods of the World Bank that pursuit of projects will depend in relation to working in the country concerned.

Introduction
Those eligible to bid, exclusions and other conditions.

Type and size of contract
Notification, advertising and pre-qualification.

Documents
Reference to bank; validity of bids, bid bonds or guarantees; conditions of contract; wording of documents; standards; use of brand names; expenditure under contract; pricing and currency of bids; payments, currency and maintenance of value; payments terms; payments advanced; price adjustment; guarantees, performance bonds and retentions; insurance; liquidated damage and bonus clauses; *force majeure*; language and disputes.

Opening of bid, evaluation and award of contract
Time allowed for preparation of bid; procedure for opening; extension of validity of bids; clarification and alteration; confidentiality; examination; postqualification; comparison of bids; local and regional preferences; rejection and award of contract.

International Development Association

Most countries which are members of the World Bank are also members of the International Development Association (IDA). Funds are provided by initial subscriptions by members, transfers from the World Bank, from accumulated income and, significantly, by additional payments by richer members. The allocation of funds depends on the wealth of the country concerned and must be for specified projects. The funds provided are described as credits. They are not loans as in the case of those provided by the World Bank as a condition of eligibility is the relatively low gross national product per capita in the country concerned. In other respects the methods of selection of projects, evaluation, bidding and supervision can be undertaken in a similar way to those used by the World Bank.

International Finance Corporation

Finance provided by the International Finance Corporation (IFC) is on a fundamentally different basis from that provided by the World Bank and IDA in that it is directed towards investment and technical support for private sector activities. It is financed by shareholders. The UK is one of the largest, and government organizations, UK firms and financial institutions have made a substantial contribution to IFC-aided projects. Technical and commercial skills in the UK, particularly in financing, have enabled developing countries with manpower and resources but, as yet, limited commercial and industrial structures, to pursue projects which would otherwise have been impossible. Financing may be in the form of equity participation, loans, debentures or a variety of other methods limited to part of the investment needed. The IFC is, however, primarily concerned with investment which is a matter for discussion between the IFC and the company concerned. Evaluation and control procedures of the type required as a part of World Bank and IDA functions are less in evidence. Application for finance is also less standardized, although basic information is needed with any application for funding. This information, which also provides a valuable guide to assessing any investment, is outlined below.

Description of project

Sponsorship and management

Business of sponsors; management and arrangements for external assistance.

Markets and sales

National, regional and export markets; projected production and market share of proposed venture; potential users and distribution; present sources of supply,

133

competition and saturation point likely; restrictions relating to products; regional integration and factors affecting market potential.

Feasibility, manpower and raw materials

Manufacturing process; technical complexities and need for special skills; equipment supply; availability of manpower and of infrastructure needed; raw material supply, costs and quality; import restrictions; plant location and plant size.

Investment needed and returns

Estimate of project cost; financial structure, equity and debt financing; profitability and return on investment and factors affecting profitability.

Government support and regulations

Government developments likely to affect projects; incentives and support available; contribution of project to national developments and government regulations likely to affect capital entry and repatriation.

Time needed for project design, preparation and completion

Regional lending agencies

There are a number of international lending agencies that operate on a regional basis and in similar ways to the World Bank group. They may work closely with the World Bank and with other supra-national bodies like the UN Conference on Trade and Development and the UN Development Programme. For those in the UK construction industry the work and objectives of the European Regional Development Fund (ERDF), the European Development Fund (EDF) and the European Investment Bank (EIB) are particularly important and are a vital part of the activities of the EEC. Others are as follows:

- Abu Dhabi Fund for Arab Economic Development.
- African Development Bank.
- African Development Fund.
- Arab Fund for Economic and Social Development.
- Asian Development Bank.
- Caribbean Development Bank.
- East African Development Bank.
- European Development Fund.
- European Investment Bank.
- Inter-American Development Bank.
- Islamic Development Bank.

- Kuwait Fund for Arab Economic Development.
- Saudi Development Fund.

Membership of a relevant regional bank or fund by a particular country is a prerequisite to benefiting from the allocation of funds and usually influences the award of contracts. Fortunately the UK is a member of most, although only recently did it become a member of the African Development Bank. Earlier it was a member of the African Development Fund only, but has supported much bilateral aid in the region, as in many others.

The activities of the Inter-American Development Bank are typical of others. It has regional members, currently totalling twenty-six, and non-regional members totalling seventeen including the UK. More than one third of the voting shares are held by the USA, but South American countries hold about 56 per cent. Since 1961 the largest allocation of funds has been for energy development, followed by agriculture and fisheries, industry and mining and transport and communications. The smallest allocation has been for urban development. It can be seen, therefore, that support for productivity and use of resources dominates while social developments, such as housing, appear to be thought largely the responsibility of national governments helped by the benefits derived from projects encouraged by multilateral aid.

Trading blocks

Regional trading blocks and associations exert their influence on local and world trade. Some developments may be partly dependent on the World Bank, regional lending agencies or trading block policies. This interdependence may be expected to continue and is seen in the UK membership of the EEC, ERDF and, on a wider basis, by the EDF. There are a number of other trading blocks also likely to influence construction markets.

Africa

In West Africa the Economic Community of West African States (ECOWAS), formed in 1975, has sixteen members whose aims are widespread. They include abolition of trade restrictions between member states, harmonization of agricultural and monetary policies, promotion of common projects in marketing, research and agriculturally based industries and the free movement of manpower, services and capital within the Community.

North America

The USA, which may also be regarded as a trading block in its own right, while not belonging to any specific block is, nevertheless, constantly in evidence in most through other means, notably by financial support for lending agencies and bilateral aid. Trading patterns may also be influenced by preferences given by the USA such as those for a Most Favoured Nation (MFN) or when there are Orderly Marketing Agreements (OMA) affecting manufactured goods.

South America

All South American countries and Mexico were members of the Latin American Free Trade Association (LAFTA), formed in 1960, whose aims were economic integration, elimination of trade and tariff barriers and the development of member states, but this has been replaced by the Latin American Integration Association (LAIA) in an attempt to achieve a better framework for trade. This involves integration promoted by partial agreements, preferential margins for intra-regional trade and regional agreements. The LAIA has three categories of membership: Argentina, Brazil and Mexico as more advanced countries; Chile, Columbia, Peru, Uruguay and Venezuela as intermediate countries and Bolivia, Ecuador and Paraguay in the least developed category, and entitled to additional concessions and technical and financial assistance from other members. Revised membership conditions permit the expansion of commercial relations with non-associated members.

Other important blocks in South America are the Basin of the River Plate Association which was formed in 1969 to encourage the economic integration and joint development of Argentina, Bolivia, Brazil, Paraguay and Uruguay and the Latin American Economic System (SELA) which was created in 1975 by twenty-five Latin American and Caribbean states. The organization is intended to complement existing agreement by economic consultation and cooperation, harmonizing the interests of the region and promoting joint and national development projects.

Eastern Europe

Construction in Eastern Europe and in some other countries can be influenced by the role of the Council for Mutual Economic Assistance (CMEA) which was founded in 1949 as a result of USSR initiatives. The member countries are the USSR, Bulgaria, Czechoslovakia, the German Democratic Republic, Hungary, Poland and Romania and, outside Europe, Mongolia and Cuba. Yugoslavia is an associate member. Apart from any other motives the CMEA was formed to encourage trade, the exchange of information and the movement of all forms of goods and services. From time to time joint projects are promoted, particularly those involving cross-frontier activities such as the movement of oil and gas through pipelines. Such projects can be influenced by national five-year plans and by the availability of funds from the International Investment Bank. Throughout, control of external currencies is a major factor. Those of CMEA countries are neither convertible nor transferable, and trade is calculated in relation to the transferable rouble. Some members of the CMEA are members of the International Monetary Fund (IMF), but CMEA does not generally adopt the type of economic external relations which is a feature of EEC activities. Foreign currency limitations encourage barter trade and switch trading by CMEA members. Such trading methods are not necessarily popular. Even when they open up new markets a new trading burden can be placed on those concerned and distort supply and pricing patterns. For those involved in world construction markets new problems can develop if the developing countries pursue barter and counter-trading methods as a means of purchasing goods and services by using their natural resources.

Middle East

To Middle East lending bodies already mentioned must be added those who encourage economic cooperation such as the Arab League, the Council of Arab Economic Unity and the Arab Monetary Fund. The Organization of Arab Petroleum Exporting Countries, established in 1968, which safeguards the interests of its nine members and coordinates the activities of their petroleum industries has an important role in construction, particularly petro-chemical developments.

South-East Asia

Asian construction can be influenced by the Association of South-East Asian Nations (ASEAN). This was established in 1967 by Indonesia, Malaysia, the Philippines, Singapore and Thailand to promote collaboration in economic, technical, scientific and administrative matters. In 1980 the EEC and ASEAN signed an economic collaboration agreement. The economic purpose of the ASEAN agreement and, indeed, the objectives of most regional blocks, can be exemplified by a further agreement signed in 1977 which gave a preference margin to ASEAN-located firms tendering for ASEAN government contracts provided at least 50 per cent of the FOB value of supplies is attributed to manufacturers within a member country.

Employment of manpower in developing countries

It is likely that developing countries will have to find employment for more than 500 m. more people within the next twenty years. This could influence the growth of construction in markets outside Europe and the likely share of the market by designers, main contractors, specialist contractors, sub-contractors and manufacturers in the UK and elsewhere, notably in industrial countries. Any country with a rapidly increasing population will require the social, commercial and industrial infrastructure to support this. However, those with the most rapidly increasing population may be least able to afford the construction needed and may not have the trade or management skills necessary. The role of international lending agencies will, therefore, be increasingly important and it is likely that they and such bodies as the UN, working with regional associations,

137

will need to establish and support programmes to meet employment and population needs. As a result there could be a significant change in the scale and type of many contracts throughout the world compared with those of recent years. Clearly there will still be the need for substantial projects, but there is also likely to be a need for many much smaller projects with a greater flexibility and in keeping with financial limits which could be imposed.

The immediate outlook for the methods of obtaining overseas construction contracts appears increasingly uncertain compared with recent years. Changes of output in a number of areas appears to be leading recently formed local companies who have gained experience to compete internationally. However, if unskilled manpower is to be employed by such companies they may still need the support of commercial, technical and management skills of UK and other European construction organizations of all sizes, particularly when complex financing and advanced technology is involved. Much is also likely to depend on the need for training designers, craftsmen and managers in the construction industry. Support for this by UK organizations is already a factor in obtaining contracts and may be more essential in the future.

Chapter 18

European Economic Community. Structure, purpose and commercial influence

Trading influence and external relations

Since it was formed it 1957 the EEC has had a growing influence on its original six members and on those, including the UK, who joined later. The UK became a member of the Community in 1973. The current ten members are Belgium, Denmark, France, The Federal German Republic, Greece, Italy, Ireland, Luxembourg, The Netherlands and the UK. Spain and Portugal and others may be expected to join. Policies pursued by the Community and the work of the Commission of the EEC and other institutions influence the UK government, local authorities and trading patterns and have a growing impact on world trade and ways in which members are involved and benefit from this. Construction in the UK, elsewhere in Europe and throughout the world is no less affected than other undertakings and industries. This influence could grow rather than diminish.

The Community has direct diplomatic contacts with over a hundred countries. It also has links with other economic groups and formalized agreements, such as observer status in the UN and in other organizations where it can sometimes exert a strong influence. Through aid programmes the Community helps developing countries. Trade between the European Free Trade Association and the Community which is about 20 per cent of Community trade encourages cooperation between the two organizations on such matters as research, the environment, fisheries, steel and transport. Association agreements exist with Cyprus, Malta and Turkey and a preferential trade agreement with Spain which could become the eleventh full member of the Community in due course. There are periodic meetings with the Council for Mutual Economic Assistance, but working relationships are limited. Recently the Community sought closer links with Turkey, Yugoslavia and the Gulf states, which led to a long-awaited cooperation agreement between the Community and Yugoslavia. A constitutional and commercial agreement has also been signed with Romania giving a new joint committee powers to develop and monitor bilateral activities.

Basis of the Community. The EEC Treaty

There are three European communities which share the same institutions. The

EEC was established in 1957 by France, Germany, Belgium, Luxembourg, Italy and the Netherlands, through the signing of the EEC Treaty. It followed from the European Coal and Steel Community (ECSC) set up in 1951. Both work with the European Atomic Energy Community (Euratom), established under the EEC Treaty to promote peaceful uses of nuclear energy. Aims of the Community are to integrate member states by economic, monetary and, eventually, political union and through the removal of tariff barriers, promotion of technology and the more efficient use of resources in agriculture and industry. Articles of the EEC Treaty govern aims and methods of working. Examples of these which influence construction are: Article 48 which provides for the free movement of manpower – local UK employment offices can provide information on living and working conditions in member states of the Community; Article 54 established the ECSC and permits the Commission to assist the financing of investment programmes, works and installations which contribute directly to the competitiveness of coal and steel producers; Articles 85 and 86 which prohibit trading agreements and practices which are deemed to be in restraint of trade, and Article 100 relating to the removal of barriers to trade. The latter has a major influence on construction markets, particularly on the standardization of products, legislation and regulations affecting health, safety and welfare.

Structure and budget

The EEC budget is financed largely by direct contributions by member states and from duties and taxes on imports from outside the Community, but also from the proceeds of value added tax (VAT) levied at a maximum rate of 1 per cent on a harmonized list of goods and services. In 1981 the total budget was estimated at almost 23 bn European Units of Account, about £11 bn. The largest contributors were Germany, the UK and France.

Administration of the Community is the responsibility of the Commission and the Council who give impetus to decision-making. The Commission makes policy proposals, but the Council makes final decisions after consulting the European Parliament, and the Economic and Social Committee which is an advisory body with nearly 150 members representing employers' organizations, trades unions and others which the Commission and the Council must consult on all major proposals. The committee give advice on its own initiative. Council meetings are prepared by the Committee of Permanent Representatives. The Court of Justice is the ultimate court of appeal.

The Commission of the European Communities

Membership of the Commission may not include more than two members having the same nationality from each member state. The number of members may, however, be amended by the unanimous vote of the Council of Ministers. Members are appointed by the governments of the member states for a four-year renewable term, during which the president and five vice-presidents hold office for a two-year renewable term. The Commission offices are in Brussels, but there is also a London office which apart from administrative and diplomatic roles can also give valuable guidance on a EEC institutions, legislation, regulations, recommendations, procedures, work being done and contacts in Brussels and elsewhere. Various Commission documents are used when drafting proposals which will affect member states and industries. A preliminary draft is produced and worked on by the Commission and may appear in any number of revisions. The next stage is a Commission proposal. This is the final version of the draft which is submitted to the Council of Ministers, which must also be referred to the European Parliament. It, too, may undergo revision as a result of representations by member states at council meetings and may be referred back to the Commission for further elaboration. The draft, after approval, becomes a directive which is, unless otherwise stated, a statutory instrument of the Council and becomes binding on member states.

The Council of Ministers

Ensuring the coordination of the economic policies of member states and taking decisions necessary for carrying out the treaties are functions of the Council. It is composed of one representative from each member state, but in practice more than one minister from each state may be present at meetings depending on the subjects being discussed. The office of president is held for six-month terms by each member country. Although under the EEC Treaty Council conclusions can be reached by majority voting discussion usually continues until unanimity is reached, except for budget and agricultural management decision. This may change as enlargement could lead to a wider use of majority voting. National votes are weighted giving ten votes to the UK, France, Germany and Italy, five votes to Belgium, Greece and the Netherlands, three votes to Denmark and Ireland and two votes to Luxembourg, total of sixty-three votes of which forty-five are needed for a majority in favour of a Commission proposal or when cast by six members for other proposals.

The European Parliament

The main task of the European Parliament is to monitor the work of the Commission and the Council, and it must be consulted on most Commission proposals before the Council may take a decision. It is composed of 410 elected members.

The Court of Justice

This is a supreme court of nine independent judges assisted by four advocates-general, all appointed jointed by EEC member states for renewable six-year tems. On all legal questions arising from Community treaties the Court is the final arbiter. It deals with disputes between member states, between member states and Community institutions and between Community institutions and firms, individuals or Community officials. It can also give preliminary rulings on the interpretation of Community law.

European Investment Bank

Internal and external investment is influenced by the European Investment Bank (EIB) which grants loans and guarantees on a non-profit-making basis to facilitate the financing of projects for developing selected regions of the Community and for modernizing or creating new activities. Each member state subscribes to and is a member of the bank. The largest are France, Germany and the UK, each of whom subscribes the same amount. To finance lending operations, the bank raises most of its funds by borrowing on capital markets. It may, however, call on member states to pay parts of any uncalled balances of capital.

The bank is authorized to lend to countries outside the Community, and has provided funds for developments in several Mediterranean countries which have association or cooperation agreements with the Community and for a number of African, Caribbean and Pacific (ACP) countries which signed the Lomé Conventions. Other such loans may be expected and the role of the European Development Fund (EDF) will be significant. Clearly, some loans will relate to construction.

Global loans

Apart from the financing methods described the EIB operates a global loan system, whereby it makes available a block sum of finance to approved intermediary institutions in Community member countries, which is then lent again for small or medium-scale ventures eligible for EIB help. The UK has benefited. By using intermediaries such as the Department of Industry in England, the Scottish Economic Planning Department, the Welsh Office Industry Department and the Northern Ireland Department of Commerce the bank is able to assist small firms which could not otherwise benefit from its resources. For instance, the bottom limit under the global system is 25 000 EUA an amount which would not be a practical economic proposition for the bank to handle directly owing to the work involved. Most financial allocations from global loans go to industrial enterprises, but under the terms of some loans it is possible to extend finance to other sectors, such as agriculture, tourism and other services. Global loans are limited to assisting ventures where the new fixed investment costs for which finance is sought will not exceed 12 m. EUA and when the promoter's total fixed assets do not exceed 30 m. EUA. Within these general conditions the EIB gives permanent priority to firms employing less than 500 people with no more than one third of the capital held by a larger company. Owing to exchange risk problems, EIB facilities were not fully developed in the UK until 1977 when the British government introduced a scheme to provide exchange risk cover. The

UK arrangement is still not strictly a global loan, but an agency agreement. However, the basic aim and end result are the same, and are designed mainly for employment-creating investment in special development and development areas and in Northern Ireland.

Business Cooperation Centre

An example of a Commission service able to assist commerce and industry is the Business Cooperation Centre. It works as an autonomous service to give every guarantee of independence and discretion to those who ask for its help and provides legal, financial and administrative information affecting cooperation and mergers between companies in member states, but will not become involved in the choice or method of cooperation in order to avoid infringing the role of professions specializing in advising businesses. The bureau meets the needs of small- and medium-size businesses particularly. However, it does not exclude any enterprises within the EEC, whatever their size or activity, from using its services. Services are free and confidential.

European Monetary System

To stabilize exchange rates between currencies and establish a better functioning of the industrial and agricultural common markets the European Monetary System (EMS) was established in 1979. All EEC members except Greece and the UK, which chose to maintain exchange rate stability outside a formal link, take part in this. The EMS provides for exchange rate movements to be monitored and contained within prescribed limits. For each currency participating in this system a central rate is fixed in terms of the European Currency Unit (ECU) of which sterling is a part. An ECU is a monetary unit based on a basket of EEC currencies identical to those used to calculate the European Unit of Account (EUA), and weighted according to GNP and volume of intra-European trade. The composition of the basket is reviewed every five years or when the weighting of one currency has changed by 25 per cent.

When a central rate is fixed for each participating currency these are used to establish a grid of bilateral exchange rates. A notional central rate has been assigned to sterling. Under the system, fluctuation margins of plus or minus 2.25 per cent from the central rate have been established for the currencies of Belgium, Denmark, France, Germany, Ireland, Luxembourg, and the Netherlands and of plus or minus 6 per cent for the Italian lira. Should the UK join the system this latter rate could also apply to the UK. Greece, while having signed the agreement of the EMS is not a member, and the drachma is not included when calculating the value of the ECU. Margins set the limits at which official exchange market intervention is obligatory, although intervention within the margins is not excluded. In addition to the grid of central rates there is a divergence indicator to give an early warning when a currency diverges in its development from the average of the others.

The EMS incorporates and expands three earlier Community credit facilities: the short-term financing and the short-term monetary support, which are the responsibility of the central banks, and the medium-term financial assistance which may be granted by the Council of Ministers to any member country with threatened or actual balance of payments difficulties. There are limitations on the UK benefiting from these credit facilities while not a full member. The system includes measures for the granting of interest-rate subsidies for loans made available to less prosperous countries from Community sources and the EIB. These loans may mount up to 1 000 EUA annually for a period of five years for the financing of investment projects which are in line with Community policy in the energy, industry and infrastructure sectors, but again the UK cannot as yet benefit.

A European Council Resolution of December 1978 envisaged the establishment of a European Monetary Fund (EMF) to consolidate the EMS, and there were hopes that this would be introduced in 1981. Until its introduction the European Monetary Cooperation Fund is empowered to receive monetary reserves from central banks and to issue ECUs against such assets. All Community countries, including the UK but not yet Greece, deposited 20 per cent of their reserves of gold and 20 per cent of dollars with the fund. These deposits are on a swap basis under national control and arrangements depend on mutual consent which implies a temporary system and role of the ECU. The creation of the EMF would provide a firmer basis for the ECU and establish an important stage in its development as a reserve currency within the Community. Although the Commission has prepared studies on the implication of the EMF there is political resistance by member states which prevents progress.

Community policies and regional development

Construction can be strongly influenced by EEC regional developments. For these the European Regional Development Fund has key policy-making and funding roles. Italy, the UK and France are currently the largest beneficiaries from the fund. Grants can contribute about half the cost of the investment assistance given by a government to qualifying industrial projects, while those for infrastructure projects are passed on in full to the local and other public authorities concerned. Applications for grants are usually prepared for a three-year period and are sent to the Commission. European Regional Development Fund allocations may also be related to those provided through the European Agricultural Guidance and Guarantee Fund. Community and regional policies on such matters as energy and environmental and similar planning have a direct bearing on construction and progress, and innovation, and expertise obtained can have applications worldwide.

Commercial policies

Harmonization of company law

All industries are likely to be affected by the EEC programme to harmonize company law as a part of the freedom of establishment of companies in member countries. Five directives have been published and a further three are being considered. The first directive, 68/151/EEC, sets out requirements for the publication of accounts, and restricts the grounds on which a court can declare a company null and void. It provides that a company is liable to third persons for the acts of its directors and agents, even if the acts are outside the objects of the company. This required a change in English law. It also made it necessary for companies to insert details of their registered number and country of incorporation on letter headings and other forms.

A second directive, 77/91/EEC, came into effect at the end of 1978. Its aims are to regulate the formation of public companies and the maintenance and alteration of their capital. Public companies now require a minimum subscribed capital of 25 000 EUA. Other changes concern the responsibility for preincorporation liabilities, a virtual prohibition on the issue of shares at a discount and stringent controls on the issue of shares for a consideration other than cash. The payment of dividends and interim dividends are also subject to some restriction.

The third directive, 78/855/EEC, is concerned with harmonization of national rules affecting domestic mergers of public limited liability companies, and when one company acquires all the assets and liabilities of another and the latter is dissolved without liquidation. The provisions define what is meant by a merger, stipulate those companies which may be merged, lay down minimum requirements for the contents, publication and supervision of the draft terms of mergers to be drawn-up by administrative or management bodies and determine the powers of general meetings and the rights of individual shareholders and of minority shareholders. Other provisions are concerned with the protection of creditors, particularly debenture-holders.

Harmonization of the annual accounts of public and private limited companies is provided for in a fourth directive, 78/660/EEC. Presentation of accounts have to conform to prescribed layouts. Assets and liabilities must be set out in accordance with fixed rules and annual accounts audited. More than a million companies in the Community will be affected. Owing to the difficulties of determining the extent to which a company must comply with the requirements of the directive to publish particulars without reference to its size, the legislation distinguishes between small, medium-size and large enterprises. For small- and medium-sized undertakings there are a number of less stringent requirements.

Proposals for future directives include a draft concerned with employee participation and company structure in the Community. Another directive, 80/390/EEC, already agreed, applies when shares are submitted for official quotation on

145

a stock exchange, not when they are issued. A seventh directive could complement the fourth directive by ensuring comparable accounting procedures for groups of companies, whether national or international, and affect multinational companies with registered offices in the Community. They would have to publish group accounts relating to all their subsidiaries throughout the world and make clear their relationships and activities within the group. Those with head offices outside the Community would be subject to the same rules applicable to companies established within the Community. The latest draft directive, the eighth, aims to harmonize the minimum qualifications required in Community countries to audit the annual accounts of limited liability companies.

Credit institutions

A directive, 77/780/EEC, published towards the end of 1977 required EEC member states to set-up licensing and authorizing procedures by 1979 to co-ordinate laws and regulations relating to credit institutions. This includes building societies. Apart from the impact of the directive on banking, insurance and other insurance and other institutions, it seems that the UK building society movement could also be affected and, in turn, construction activities. Member states are permitted under the directive to delay implementing the proposals for up to eight years. Progress will, therefore, depend on building society impetus with that of other institutions and market forces. Much will also depend on how far the Commission insists on the directive being implemented. The special nature of some lending authorities in Denmark, Ireland, Luxembourg, the Netherlands and the UK has been recognized and a list of exemptions was published in October 1978. In the UK, application of the directive to trustee savings banks and building societies may be deferred.

Export insurance and credit

The EEC has a common customs tariff for imports, but exports are not subject to such regulation. This has concerned the Commission for a long time. Many countries have introduced export credit systems to protect their exporters from certain risks and the Commission has said that the great divergence that exists in the export insurance and credit systems of member states leads to unfair competition. In an attempt to bring the situation under some control, a draft directive laying down uniform principles to be applied by member states in their systems of credit insurance and export credit guarantees has been prepared. The proposals seek to eliminate differences between the external credit systems of member states and in third country markets, to facilitate cooperation between undertakings of different member states in the search for external markets and set out principles for medium- and long-term transactions with public and private buyers while leaving it to local insurers and guarantors to draw up policies in accordance with requirements of national laws based on these principles.

There is little conflict with UK procedures already operating. Before the policies are adopted, however, there must be consultations with the Commission and other insurers and guarantors. Three main types of financing of export transactions liable to be insured or guaranteed are covered, which are:

Supplier credit
This is the traditional source of export credit. An exporter concludes a contract with a buyer in a third country in which it is undertaken to supply goods or services and agreed that the goods are to be paid for on a credit basis over a specific period. The exporter covers against the risk of non-payment by taking out credit insurance. The directive sets out the general principles to be applied concerning the scope of the guarantee, events constituting a cause of loss and the period before ascertainment of loss, the effective date and the extent of the guarantee, indemnification and premiums.

Bank guarantee
To finance supplier credits granted to a buyer an exporter normally has recourse to a bank. The bank is not a party to the export contract and its function is limited to providing the necessary financing without risk of loss. Accordingly, the bank seeks from a credit insurance undertaking a non-conditional 100 per cent guarantee against default in payment of the credit granted. The principles applicable to these guarantees are also covered by the directive.

Financial credit
The use of financial credit or purchaser credit increased in recent years. Under this the export transaction is financed by a loan granted by the exporter's bank directly to the buyer or to the buyer's bank. The only export credit insurance required is in respect of this loan, and it is granted directly to the bank and not to the exporter. Principles governing this form of credit insurance are also stated in the directive.

Complete harmonization of premiums is not envisaged at present, but in order that systems of premiums may be made clearer the directive provides that the rules, scales of premiums or premium tariffs applied to the export credit insurance or guarantees required by the directive will be notified to the Commission and to insurers. The Commission intends to continue to seek a harmonized system of export credit insurance premiums in which the premiums collected are adequate to cover losses over a number of years. Under the directive a consultative committee on credit insurance and export credit guarantees consisting of representatives of member states and the Commission would be established. Three years after the implementation of the uniform principles, having consulted this committee, the Commission will present a report on the experience gained. It could, as a result, make proposals to adjust the uniform principles to meet practical requirements.

Export financing

Commission policies to ensure fair competition, including that for exporting, can be seen from action which was taken in 1979 in relation to an export financing scheme in France. Under a special scheme for investments to increase exporting firms' production capacity specialized financial institutions floated loans on foreign financial markets in foreign currencies with an exchange rate guarantee provided by the state. The interest rate on the loans was lower than that which could be obtained for loans in French francs on the domestic market and the funds obtained were again lent to French firms in the form of long-term loans in French francs at a rate of interest approximately 2 per cent lower than the terms given by institutions for their own normal loan business. Loans covered up to 35 per cent of investment costs intended to increase production capacity and the firms concerned undertook to increase the proportion of their sales accounted for by their exports to non-member countries. This favourable rate of interest was charged by the financial institutions in question only because the state guaranteed against the exchange risks they ran when converting funds raised in foreign currencies into loans in French francs. Without the guarantee they would either have had to exclude the conversion operation or cover themselves against the risk by charging firms interest rates or a premium which would have brought the cost of the loans up to ordinary French commercial rates.

Provision of a state guarantee of any type constitutes an aid, whether or not the state concerned must later make good a loss. The scheme favoured certain undertakings and the production of certain goods since it applied only to investments that expanded production capacity in firms undertaking to increase their sales in non-member country markets, even if the output as a result of the investment was not used for the purpose of such sales. The Commission considered that such an aid scheme affected trade between member states and threatened to distort competition. As a result France was required to take measures to ensure that loans given by specialized financial institutions under the special financing scheme for investment by exporting firms were granted only when certain conditions were fulfilled. These were that when such loans were granted under programmes for an entire sector or branch of industry, such programmes should be notified to the Commission in accordance with Article 93 (3) of the EEC Treaty and, when such loans were granted to one or more individual firms, significant cases should be notified to the Commission.

Energy

Of all developments in recent years the availability of energy and its rising costs would have created a need for European countries to establish common policies. Throughout the EEC there is clearly great concern about energy shortages, prices and ways in which Community progresss will be affected in the short and long term. Again, construction is a key factor in any planning. In 1978 at the Bremen European Council meeting the Commission emphasized the need for the appraisal and coordination of energy programmes and, as a result a programme was adopted for the four year period starting on 1 July 1979. The upper limit for expenditure was put at 125 m. EUA. The Commission is responsible for the implementation of the programme assisted by advisory committees on programme management set up for this purpose and is to keep the Scientific and Technical Research Committee informed. Work is being carried out under contract. The sectors to be considered in the programme are the following.

Energy conservation

Residential and commercial applications; industry; transport; energy transformation and energy storage.

Production and utilization of hydrogen

Thermochemical production; electrolytic production and transportation, storage and utilization.

Solar energy

Solar energy applications to dwellings; thermomechanical solar power plants; photovoltaic power generation; photochemical, photoelectrochemical and photobiological processes; energy from biomass; solar radiation data; wind energy and solar energy in agriculture and industry.

Geothermal energy

Integrated geological, geophysical and geochemical investigations in selected areas; subsurface problems of natural hydrothermal resources; surface problems related to the use of hydrothermal resources and hot dry rocks.

Energy systems analysis and strategy studies

Improvement and further development of the medium- and long-term energy conservation energy models; development of new concepts for energy systems representation; development of new means for better communication between model builders and model users and world energy modelling.

The Commission has emphasized that energy research, development and planning should be an international venture and that the EEC should contribute to the work of the International Energy Agency and other international bodies. Most energy research programmes are indirect, and involve those who are under no obligation to make their work known to international organizations or other countries. A feature of EEC planning was to limit oil consumption in 1979 to

Table 18.1 Energy production and imports 1977–79 (m. tonnes of oil equivalent)

	1977		1978		1979	
	Production	*Net imports*	*Production*	*Net imports*	*Production*	*Net imports*
Solid fuels	174.3	27.7	173.5	26.3	174.7	31.6
Oil	48.6	480.3	63.0	472.0	95.0	450.0
Natural gas	142.2	16.9	135.2	30.6	137.9	38.6
Primary electricity and other types	61.5	3.8	60.3	3.3	64.4	2.7
Total	426.6	528.7	432.0	532.2	472.0	522.9

Import targets for oil (m. tonnes)

	1980	1985
Belgium	30	29
Denmark	16.5	11
France	117	111
Germany	143	142
Ireland	6.5	8
Italy	103.5	124
Luxembourg	1.5	2
Netherlands	42	50
United Kingdom	12	—

Regulations for financial support for energy saving and exploiting alternative sources

79/725	–	Demonstration projects
79/726	–	Alternative energy sources
79/727	–	Solar energy
79/728	–	Solid fuel liquefaction

500 m. tonnes and to put a target for imports in 1985 at 1978 levels. Details of energy production and imports during 1977–79, target oil imports and key regulations for financial support for energy saving and exploitation are given in Table 18.1.

Apart from preparing directives designed to encourage energy saving, recommendations and resolutions the Commission has also approved a resolution intended to reduce the ratio between the rate of growth in gross primary energy consumptions and the rate of growth in GNPs progressively below 0.7 by 1990. It was also agreed that in 1980 member states should adopt energy saving programmes, including energy pricing practices, based on guidelines adapted to national priorities and conditions. Those set out in an outline programme recommended to every member state of the Community cover:

Energy pricing practices

- Taxes on energy should be maintained or increased to reflect the scarcity of energy as a factor of production.
- Prices should be linked to the costs of replacing and developing resources.
- Energy prices in the market should be subject to the greatest possible inspection.
- Publicity about prices, and the costs and consumption of equipment using energy, should be developed as widely as possible.
- Specific measures to encourage the rational use of energy.

Energy saving in the home

- A substantial upward revision in mandatory thermal performance requirements for new buildings and heating systems.
- Regulations to ensure individual metering and billing and control of heating systems in multi-occupied buildings.
- Performance standards and control of servicing of heating systems.
- Publicity campaigns and advice centres for energy saving in the home.
- Financial aid and a programme for dwellings in public ownership.
- Labelling to indicate the energy consumption of domestic appliances.

Energy saving in industry

- Requirements for energy audits, especially in industries consuming large volumes of energy.
- Financial aid for advice and expertise for small- and medium-sized businesses and publicity campaigns.
- Financial aid and tax credits to support investment to save energy.
- Financial aid to promote new technologies, equipment and designs.

Energy saving in offices and commerce

- A public sector programme.
- Mandatory building codes for new offices.
- Performance standards and control of servicing of heating, cooling and ventilation systems,

Energy saving in transport

- Information and publicity campaigns.
- Standard tests of the efficiency of fuel use, and publicity.
- Discussions with industry on voluntary targets for the efficiency of fuel use of new cars.

Energy production

- Measures to encourage the rational production and use of heat and power.

Information and education

- Sustained publicity about energy saving.
- Education programmes in schools, technical colleges and universities and for professional training.

Research and development.

Investment and restructuring

All EEC work seems to be directed towards change and will therefore have a bearing on international activities of all types. Energy saving is clearly likely to lead to investment requirements and potential markets for construction, particularly of power stations other than oil burning. The promotion of investment generally is one of the EEC priorities. As an example, following EEC Council decision 78/870/EEC, the Commission is empowered to contract loans to promote investment within the EEC. Borrowings are to be used for loans to finance investment projects for infrastructure and energy. The first tranche was for transport, telecommunications, agriculture improvements, water supply works and environmental protection. The Commission decides whether or not projects are eligible and contribute to resolving problems and reducing regional imbalances and unemployment. This type of activity must be seen and understood in relation to purpose of the Community, its institutions and opinions, resolution, regulations and directives and, in turn, ways in which these are implemented by national governments thereby influencing home and international trading patterns.

Chapter 19

European Economic Community. Construction industry planning

A part of a broader approach

Construction industry progress within the EEC and its role in world markets cannot be separated from EEC planning and the structure, purpose and influences of the Community as a whole. Few in the construction industries of member states have been satisfied with home demand. In most countries all sectors of construction are on a plateau of output, and high energy costs and changed patterns of orders create further uncertainties. Optimism about opportunities for EEC organizations in overseas construction markets which was evident a few years ago is now far less, and further reduces confidence. Overseas markets are extremely competitive and are being pursued by contractors, suppliers and others from Eastern Europe, Asia and the Middle East and North and South America in ways which were not in existence ten years ago. Many UK and other EEC organizations may be expected to compete and succeed in world construction markets, but EEC planning may have a growing part to play in such competition, including measures to maintain output and employment at home.

Sound investment programmes are also needed. The support of the European Investment Bank (EIB), aid to non-associated countries and that made through the European Development Fund (EDF) and communications with new members of the Community require the most detailed attention. The Commission is giving much thought to energy, infrastructure, transport and communications planning for which construction industry support is certain to be required. Participation in projects financed by the EDF can benefit the UK construction industry and, to some extent, the UK balance of payments.

The general progress of the economy throughout the Community must influence construction output. Continually rising prices of oil, recession factors and unresolved problems in member countries led to poor EEC growth in recent years. While the Community can aim for a growth of the GDP in the majority of states, the weaker performance of others still results in an overall poor average. Slower growth and a continued rise in the population of working age, means that unemployment in most member countries is expected to rise. Higher energy costs also lead to rises of consumer prices and inflation generally. The EEC balance of payments forecast is also depressing. Although the Commission hoped for an improvement by the end of 1981, balance of payments deficits arose compared with earlier surpluses.

Apart from these overall industry factors there are a number of EEC matters which have a direct impact on construction output and many associated activities. Until a few years ago the Commission of the EEC appeared to have given little thought to the construction industries of member states. No doubt other matters gave more cause for attention, but it is now recognized that construction is an essential part of EEC development.

In 1972 the Commission approved procedures involving the exchange of views between experts from member states and Commission departments to identify projects likely to encourage a unified EEC construction market. Early in 1975 proposals to strengthen this planning were also agreed by the Commission. Short-term objectives included harmonization of units of measurement and of data presentation; improvement and development of the statistics for public sector construction and engineering, particularly those concerning the erection of residential and non-residential building; promotion of comparable national forecasts of output and proposals for harmonizing the technical basis of building legislation in EEC member states in relation to fire protection and structural stability. These activities largely involved work by government representatives, but some industry representatives were consulted. Objectives given for supporting work included the preparation of a glossary of the main technical terms used in national laws and regulations relating to construction; an inventory of essential national legal regulations and administrative systems; a report on the functional requirements and basic principles that should be taken into account in structural safety inspections and for fire prevention and preparation of procedures for the technical approval of construction materials and methods. Work was also started on dimensional coordination and thermal insulation. Longer-term activities were to cover technical requirements forming the basis of all building regulations and ways in which performance may be specified.

Public sector contracts

One of the first EEC directives to have a direct control over a construction activity was that providing for the advertisement of public sector contracts in the Official Journal of the European Communities. Directive 71/305/EEC gives conditions for this and the threshold, currently 1 m. EUA, at which advertisement must start. The rates used earlier for converting units of account into national currencies were defined by referring to the gold parity unit of account. However, the conversion of the threshold into national currencies on the basis of this definition failed to reflect proper values and produced varying effects in member states. Article 10 of a financial regulation of 21 December 1977 defined a EUA as representing the average variation of the value of the currencies of member states, whereas the exchange value of this unit of account in each of the currencies concerned is determined daily. Following a further Council directive, 78/669/EEC, directive 71/305/EEC was amended. As a result the exchange value in a national currency for determining the value of the threshold amount specified in the directive is the average of the daily values of the currency over the preceding twelve months, calculated on the last day of October every two years, to take effect from the following January. This exchange value is published in the Official Journal during the first part of November and governs when UK local authority works contracts must be advertised in the Official Journal. The cost in sterling has ranged between about £520 000 to £670 000. There is also a directive, 77/62/EEC, covering public sector supply contracts.

Harmonization

Harmonization is a feature of all Commission planning affecting commercial and other procedures and matters specific to industries. A variety of proposals for directives covering construction activities and products is being considered, and are listed below. Their potential impact on exports cannot be ignored.

Electrical

Plugs and socket.

Lifts and lifting apparatus

Certification and marking of wire ropes, chains and hooks; common provisions for lifting and mechanical handling appliances; tower cranes for building work and safety aspects; builders' hoists; conveyor belts; electrically operated lifts and powered industrial trucks.

Oil pipelines

Oil pipelines (construction).

Gas appliances

Framework directive for gas appliances and associated safety and control devices; instantaneous production of hot water for sanitary purposes; central heating boilers; gas governors and independent space heaters.

Gas pipelines

Gas pipelines (construction).

Construction materials

Cement; ceramic tiles; flat glass; PVC pipes; fire resistance; water supply equipment; construction products (framework directive) and safety of glass products intended for use in buildings.

Building and civil engineering equipment and machines

General directive; determination of the noise produced by construction plant and equipment; permissible sound levels for pneumatic concrete-breakers and jackhammers; permissible sound emission levels for tower cranes, grass cutting machines and generators for welding power supply; noise emitted by air compressors; permissible sound emission levels for bulldozers, loaders and excavators; site safety; roll-over protective structures; falling-object protective structures; dimensions of operators' cabins; protecting guards and shields and road safety.

Metal scaffolding

Working metal scaffolds.

Consumer protection

Approximation of the laws of member states on liability for defective products and consumer protection in respect of contracts negotiated away from business premises.

Statistics

Earlier planning gave rise to a directive, 78/166/EEC, on statistics. As a result, from 1979 facts are available about EEC construction based on national figures. The directive requires member states to collect statistical data about cycles and trends to a specified pattern. The definition of activities is given in the General Industrial Classification of Economic Activities Within the European Communities (NACE) published in 1970. Surveys carried out must relate to undertakings employing twenty or more persons. It also relates to building permits. The data to be provided by member states is as follows.

Building permits
- Number of permits granted for residential buildings with an indication of the number of dwellings and the habitable floor space and/or volume to be constructed.
- Number of permits granted for non-residential buildings with an indication of the number of buildings and the useful floor space and/or the volume to be constructed.

Indices of industrial production
- Index of production for building and civil engineering.

156

- Index of production for building.
- Index of production for civil engineering.

Orders received
- Index of orders received for building and civil engineering.
- Orders received for building in value or value of residential and non-residential buildings started.
- Orders received for civil engineering, in value.

Other data
- Number of employees, specifying the number of manual workers.
- Gross wages and salaries.
- Volume of work done.
- Number of hours worked in building.
- Number of hours worked in civil engineering.

Construction products. Framework directive

Export of construction products is an essential industry of all EEC member states. Commission plans to achieve a degree of common standards and procedures for construction products, known as the construction products framework directive has, therefore, wider implications than those for Community construction alone. Although not yet approved, proposals should also be seen in relation to international construction markets. The proposals are a part of the Commission of the EEC programme for the removal of technical barriers to trade. They arose, among other reasons, because the procedures for removing barriers under Article 100 of the EEC Treaty were said to have fallen behind. Apart from dealing with complex technical matters, the method of harmonization proposed would rely on implementing derectives adopted by the Commission with the approval of a qualified majority of a Committee of Implementing Directives composed of representatives of member states rather than by the unanimous decision of the Council which is current procedure. Although similar committees have been established in the past they were for restricted purposes. The proposals introduce a wider use of a procedure which could lead to a movement of power away from the Council and from national to Community institutions.

Implementing directives are intended to ensure that products must be so manufactured and dimensioned that they can be used for their intended purposes, that they are stable in use, that their durability is reasonably predictable and that the proportion of defective products remains within acceptable limits. Nearly thirty kinds of products to which priority is to be given in the presentation and adoption of implementing directives could be affected. They include partitioning systems, floor components, windows, doors, plastic and synthetic floor coverings and wall claddings, sanitary equipment and pipes and valves. Three possible procedures for certifying that products conform to Community standards are given: EEC-type approval by member states, EEC-type examination carried out by bodies authorized for this purpose by the member states and EEC self-certification by the manufacturer acting on his own responsibility. Provision is made for optional harmonization and total harmonization. If an implementing directive prescribes the former then products which conform to its provisions and which bear the appropriate EEC mark may be marketed and used within the Community without restriction. The marketing and use of products which do not conform remains within the discretion of member states. However, total harmonization means that member states are obliged to prohibit the marketing of products covered by the directive which are not in conformity with its provisions.

Article 155 of the EEC Treaty empowers the Commission to exercise powers for the implementation of rules laid down by the Council. Its use in relation to the draft directive is a new procedure. With the enlargement of the Community and possibility that this will retard the work of the Council it may be necessary to delegate more authority to the Commission under Article 155, and there may be a good case for relieving the Council of technical decisions related to policies that it has already agreed upon provided there are suitable safeguards. Harmonization of construction products could be an appropriate subject for delegation. Council directives are published in draft when they are proposed by the Commission, but there is no publication of them by the Commission until they come into force. This means that the opportunity for public debate, other than that initiated by bodies approached at the drafting stage by the Commission, would be lost if the proposed procedures are adopted. Drafts would not come before the European parliament, neither would they be put before the parliaments of member states. There are many doubts about the introduction of the directive. Most industry suggestions are that any barriers to trade could be overcome by the application of Article 100 of the EEC Treaty and by continuing work by the Comité Européen de Normalisation (CEN) and the International Organization for Standardization (ISO).

Product liability

Another proposed directive sets out proposals to rationalize the laws of member states on product liability, and could also influence exports. The proposals place liability without fault for personal unjury or damage to personal property arising from defective movable articles on manufacturers or on importers in the case of goods imported from outside the EEC. In the proposals, liability is placed on the manufacturer for damage caused by a defect in the article, 'whether or not he knew or could have known of the defect', but the producer is not liable if he can prove that he did not put the article into circulation or that it was not defective when he did so. 'Damage' includes death or personal injury.

The amount of damages to be awarded in individual cases is not specified since this is a matter for the courts to determine in EEC member states. However, there is a maximum figure for total liability of 25 m. EUA for personal injuries caused by identical articles having the same defect. For damage to property the limit is 15 000 EUA for movable property and 50 000 EUA for immovable property. Proceedings for the recovery of damages may only take place within three years from the date when the injured party has all the necessary information to bring proceedings, and, unless proceeding have been instituted, the liability of the producer is extinguished ten years after the date the product was put into circulation. Liability extends only to movable property since all member states have special rules to cover defective buildings, but where movable objects are used in the erection of buildings, or are installed in buildings, the producer remains liable in respect of these.

Planning and output

Environment

Environmental demands have led the Commission to introduce a draft directive concerning the assessment of the effect on the environment of public and private development projects. Most EEC members have complex and sophisticated legislation to control the impact of developments, but the Commission now clearly wishes to impose its own mandatory procedures.

Earlier Commission programmes emphasized that the best policy was to prevent the creation of pollution or nuisances at source rather than by trying to counteract their effects. They said that the effects on the environment should be taken into account at the earliest possible stage in all the technical planning and decision-making, and it was necessary to evaluate the effects of any measure that is adopted or contemplated by members. This led to a Community agreement that procedures for assessing environmental impact be prepared by the Commission.

If the proposed directive is accepted priority is to be given to the introduction of assessment principles in planning procedures for the control of development projects. Planning permission or approval of development projects likely to have a significant effect on the environment would be granted only after assessment, and authorities and developers will be required to provide information on such projects and on alternatives. Apart from a wish to create common standards it also seems that the Commission thinks that disparity between the measures in force or in preparation in member states for assessing environmental effects may create unfavourable competitive conditions and, thereby, affect the functioning of the common market. They have said that it is necessary to undertake the approximation of relevant national laws under Article 100 of the EEC Treaty. Recourse may also be made to Article 235. Great importance is put on transfrontier influences and projects likely to influence frontier regions.

The criteria for assessments are extensive. The environmental sensitivity of sites is to be considered in relation to water, air, soil, landscape, flora, fauna, built-up environment; the quality of these resources; absorption and diffusion aptitude of the natural environment and regenerative capacity of resources. Sites to be considered sensitive are wetlands, coastal zones, mountain and forest areas, nature reserves and parks, areas already classified or protected under national legislation and densely populated areas.

Although some development projects are to be governed strictly by the proposed directive others, including different sectors of the same industries, may be regulated by criteria established by member states. There is resistance by some members and many in the construction industries of members to the proposals, particularly as they can duplicate existing national laws and regulations. Projects which could be governed by proposed procedures are summarized below.

Development projects affected by EEC criteria
- Extractive industry
  Extraction and briquetting of solid fuels; extraction of ores containing fissionable and fertile material and of bituminous shale and extraction and preparation of metalliferous ores.
- Energy industry
  Coke ovens; petroleum refining; production and processing of fissionable and fertile materials and generation of electricity from nuclear energy.
- Production and preliminary processing of metals
  Iron and steel industry, excluding integrated coke ovens; manufacture of steel tubes; drawing, cold rolling and cold folding of steel and production and primary processing of non-ferrous metals.
- Manufacture of non-metallic mineral products
  Manufacture of cement and asbestos-cement products.
- Chemical industry
  Petrochemical complexes for the production of olefins, olefins derivatives, bulk monomers and polymers; chemical complexes for the production of organic basic intermediates and complexes for the production of basic inorganic chemicals.
- Metals manufacture
  Foundries; forging; treatment and coating of metals and manufacture of agricultural tractors, road tractors, standard and narrow-gauge railway and tramway rolling-stock, aeroplanes and helicopters (including the engines).
- Food industry
  Manufacture and refining of sugar and manufacture of potato flour, potato and corn starch.
- Processing of rubber
  Rubber factories and manufacture of rubber tyres.
- Building and civil engineering
  Construction of highways, railways and airports and permanent motor and motorcycle racing tracks.

Development projects affected by criteria established by member states
- Agriculture
  Land reform; projects for cultivating natural areas and abandoned land;

water management projects for agriculture including drainage and irrigation, intensive livestock rearing units and forest management plans.

- Extractive industry
  Extraction of petroleum; extraction and purifying of natural gas and extraction of minerals other than metalliferous and energy-producing minerals.
- Energy industry
  Research plants for the production and processing of fissionable and fertile materials and production and distribution of electricity, gas, steam and hot water, but not the production of electricity from nuclear energy.
- Manufacture of glass fibres, glass wool and silicate wool.
- Chemical industry
  Plants for the production and treatment of fine chemicals, dyestuffs, active ingredients for the preparation of pesticides and pharmaceutical products, paint and varnishes, elastomers and paroxides and storage facilities for petrochemical or chemical products.
- Metals manufacture
  Stamping and pressing; secondary transformation treatment and coating of metals; boilermaking, manufacture of reservoirs, tanks and other sheet-metal containers; manufacture and assembly of motor vehicles, including road tractors; manufacture of motor vehicle engines and shipbuilding.
- Food industry
  Slaughterhouses and fish-meal and fish-oil factories.
- Textile, leather, wood, paper industries
  Wool washing and degreasing factories; tanning and dressing factories; manufacture of veneers, plywood, fibre boards, particle boards, pulp, paper and paper boards and cellulose mills.
- Building and civil engineering
  Large residential and commercial buildings; construction of roads, harbours, airfields; river draining and flood relief works; dams and water reservoirs; installation of pipelines for long-distance transport; incinerators for municipal waste, including animal effluents and sludge from waste water treatment; facilities for the disposal of animal carcasses and sewage treatment plants.

Article 6 of the proposed directive specifies information which must be given, if necessary in cooperation with the relevant authorities and administration. Information covers particularly the description of the proposed project, description of the environment likely to be significantly affected, an assessment of the important effects, a review of the relationship of the project with existing environmental and land-use plans and standards for the area, a justification for choosing the project among other alternatives and a non-technical summary. Government and local authorities must decide the channels best suited for giving the public complete information within a suitable time limit, for ascertaining their views and for arranging consultations. If a project is likely to have an important effect on the environment of another member state the information must be given to the state concerned so that it too has an opportunity to make comments.

Urban and transport developments

Growth of large urban concentrations within the EEC and their impact on transport needs could clearly influence transport planning and offer one of the most likely ways to expand construction output in the Community. The EEC Treaty envisaged a Community transport policy, but too little progress has been made on this. The growth of inter-Community trade has shown that national systems are not always adequate for cross-frontier transit, and it has been said within the Commission that there is a need to approve finance which could be used to assist selected projects, of which a cross-Channel link could be one. Any programme may be mainly concerned with roads, rail and waterways, but some aspects of ports and airports could also be taken into account. There has been a Commission examination of traffic flows of major Community routes and conditions which create bottlenecks. The results could enable the Commission to prepare a list of priority projects and proposals for financing them.

Energy

Future investment in energy needs could put great demands on national and EEC budgets. Construction output in this field as a result is likely to be massive in the foreseeable future compared with any other sector. Another aspect of energy planning is a proposal for a regulation for the introduction of consultation procedures in respect of power stations likely to affect the territory of another member state, particularly consultation initiated by a member which considers that a power station planned by another is likely to have effects on its territory which are not covered by Article 37 of the EEC Treaty; by the state on whose territory it is proposed to build the power station and by the Commission. The proposals seek to ensure that when consultation is requested in respect of a proposed power station the Commission may ask a state to supply data to enable an assessment to be made of the permanent, temporary or potential effects on the atmosphere, soil, surface and ground waters of the other member states; risks for neighbouring states likely to arise from any malfunction of the power station or from accidents and about installations of all types, whether existing, under construction or planned, in a neighbouring member state when the hazards and effects on the environment and health might be cumulative.

Consultation

Various procedures exist as a part of the structure of the EEC for consultation on proposed directives and other matters. Much is undertaken direct with government departments of member states. In the UK the Department of the Environment is a focal point on construction matters, but some aspects of these and many other subjects are also handled by other government departments. The Treasury, the Foreign and Commonwealth Office, the Department of Industry, the Department of Trade and the Department of Energy can have over-lapping functions on construction topics. The Commission also has a responsibility to receive comments by sectoral associations. These are EEC groupings of professional, trade or other bodies. One of the most important is the Fédération Internationale Européenne de la Construction (FIEC) through which the National Federation of Building Trades Employers and Federation of Civil Engineering Contractors may work on behalf of the UK interests. Also important for manufacturers is the work of the National Council of Building Material Producers. On a broader aspect the role of the Confederation of British Industries and chambers of commerce through their European links is also most important. In practice the Commission is most approachable and tries to listen to as much informed comment as possible, within the limits of what is practicable. Time may, however, restrict consultation only to agreed procedures, and increasingly so as the Community becomes larger.

International contributions

Whatever the reactions to the type of planning outlined, there seems little doubt that many governments will see advantages in studying EEC activities and decisions. Some will do this through traditional bilateral links, but there are others, perhaps for historical or current reasons, who may prefer to deal with supra-national bodies lie the EEC rather than with particular countries. The EEC and its institutions has, therefore, a significant and growing role in international construction.

Chapter 20

European Economic Community
European Development Fund

Aid can benefit the donor

Most industrial countries provide assistance to developing countries. There may be many reasons for this; diplomatic, political, economic, trade or as part of short- or long-term plans to help the recipient. Whatever the reason the donor can also benefit. Aid given by the EEC through the European Development Fund (EDF) is an essential part of Community policy. European Development Fund grants are channelled through provisions established by the EEC Treaty for association with overseas countries, and should be seen in relation to other lending agencies, particularly those of the region concerned.

Lomé Conventions

Most important channels for EDF aid have been the two Lomé Conventions in which African, Caribbean and Pacific states (ACP) participated. They benefited from Lomé I and should now gain from the signing of the Lomé II Convention, but so, too, can EEC consultants, construction contractors and manufacturers and suppliers. The first Convention was signed in Lomé on 28 February 1975 and came into force on 1 April 1976, although the trade provisions were applicable from 1 July 1975. It expired on 1 March 1980, and was renewed in a re-negotiated form from that date until 1 March 1985. The forty-six countries that signed Lomé I have now increased to the fifty-nine, listed in Table 20.1.

One of the objects of the Lomé II Convention is to improve and increase industrial development in the ACP countries. It lays particular stress on the need to develop new sources of energy, on prospecting and on production in ACP countries. The European Investment Bank (EIB) is expected to play an important role in investment policies, but all concerned in the Convention agree that an influx of private capital would be beneficial. This is encouraged by a general clause in the Convention on investment protection agreements guaranteeing equal treatment for all Community countries regarding bilateral agreements. Particular efforts are to be made to give technical and financial assistance to small and medium enterprises. Funds to carry out schemes specified by the Convention were increased, compared with Lomé I. A comparison is shown in Table 20.2.

Table 20.1 ACP states

Bahamas	Grenada	São Tomé Principe
Barbados	Guinea	Senegal
Benin	Guinea-Bissau	Seychelles
Botswana	Guyana	Sierra Leone
Burundi	Ivory Coast	Solomon Islands
Cameroon	Jamaica	Somalia
Cape Verde	Kenya	Sudan
Central African	Kiribati	Surinam
Republic	Lesotho	Swaziland
Chad	Liberia	Tanzania
Comoros	Madagascar	Togo
Congo	Malawi	Tonga
Djibouti	Mali	Trinidad and Tobago
Dominica	Mauritania	Tuvalu
Equatorial Guinea	Mauritius	Uganda
Ethiopia	Niger	Upper Volta
Fiji	Nigeria	Western Samoa
Gabon	Papua-New Guinea	Zaire
Gambia	Rwanda	Zambia
Ghana	St Lucia	Zimbabwe

Table 20.2 EEC financial aid to ACP countries (m EUA)

EDF		
Grants	2 145	2 928
Special loans	445	504
Risk capital		
formation aid	97	280
Stabex	380	550
Minerals	—	280
EIB	390	685
Outside the Convention	—	380
Total	3 457	5 607

The need to exploit mineral resources was emphasized in the Convention. Since 1961 there has been a deterioration of foreign investment in developing countries, particularly for mining exploration, from a total of 57 per cent of total investment to only 13.5 per cent between 1973 to 1975. Clearly reluctance to invest in minerals production hinders development, and could pose problems for Community industries. It is estimated that by 1985 Western countries will depend on developing countries for between 50 to 100 per cent of such vital minerals as cobalt, copper, phosphates, tin and tungsten. Without investment and exploration now the minerals will not be available. One of the features of the second Lomé Convention is the creation of a fund of 280 m. EUA to permit financing through a special form of loan for projects designed to maintain the

165

potential capacity of producers of minerals should they suffer from a fall of export earnings or unforeseen events.

Assistance to non-associated countries

Community aid to non-associated developing countries, which are those outside the Lomé Convention, was about 140 m. EUA in 1980. Directed towards a population of 1 250 m. in Asia, Latin America, and Africa, the Community contribution is small in relation to that needed. However, the funds are in addition to ACP aid and are a further source of finance likely to benefit construction output, albeit in a limited way. While this particular aid is directed towards the poorest countries, more prosperous countries will be helped if the projects concerned are clearly in favour. The Commission prefers to see aid for one year concentrated on a substantial project in one country rather than finance a small project each year in every eligible country.

Contracts financed by EDF

The usual procedure for placing contracts to be financed by the EDF is by open international invitation to tender, but invitation may be to a restricted list, especially when there has been pre-selection. It may also be made under an expedited procedure and, less frequently, the contract may be placed by private treaty. Open tenders ensure the widest possible participation. For this reason the Commission has, whenever possible, used this method. The main method of publicity is through the Press in the form of a notice of invitation to tender. This gives the relevant information, including the purpose of the contract, place and time, schedule for works or delivery of goods, directions for ascertaining detailed information, submission of tenders and their examination and eligibility for participation in the tender. The notice of invitation to tender is published simultaneously in the Official Journal of the European Communities and in the official gazette of the state, country or territory concerned. The types of contracts are listed below. Contracts are governed by general conditions, giving rules for their preparation, award and execution, and by special conditions for each contract.

Works contracts

- Erection of buildings.
- Communications and port installations.
- Water supply and hydraulic agricultural improvement schemes.

Supply contracts

- Materials, products and equipment for specific works contracts.

- Agricultural supplies.
- Other products such as vehicles, plant, furnishings and maintenance materials.

Service contracts

- Surveys for proposed projects including feasibility and profitability studies.
- Supervision and support for management of projects.
- Technical assistance and support for management.

Administration of the EDF is the responsibility of the Commission of the EEC through the directorate general responsible for development. There are several directorates supervising aspects of allocating funds. They cover the following.

General development policy and internal relations

- Basic planning and coordination of member state policies.
- Evaluation of aid operations.
- Aspects of commercial policy and commodities policy concerning developing countries.
- UNCTAD and international relations covered by development policy.
- The Courier (EEC–ACP).

Africa, the Caribbean and the Pacific

- West Africa.
- East Africa.
- Caribbean and Indian and Pacific Oceans.
- General questions concerning the ACP Convention and aid programming and coordination.
- Maghreb, Israel, Egypt, Jordan, Lebanon, Syria and coordination matters relating to the Mediterranean area.
- Relations with Community and EEC–ACP institutions and with non-government bodies.

Projects

- Agriculture.
- Livestock and fisheries.
- Roads.
- Industry, energy, telecommunications and general infrastructure.
- Urban works and social infrastructure.
- Water engineering.

Operations

- Food aid and exceptional aid.
- Industrial cooperation, trade promotion and regional cooperation.
- Stabilization of export earnings.
- Training.

Finance and administration

- Financing.
- Authorization of payments and accounting.
- Invitations to tender, contracts and disputes.
- European Association for Cooperation.
- Secretariat of the Financing Committee.

Throughout the assessment and allocation of funds for contracts the emphasis is on equal competition by those who are entitled to bid from member states. There can, however, be the advantages of familiarity with some countries and procedures as a result of earlier associations, as in the case with the UK in some cases and with France in others. No other members of the EEC have such widespread former links, although most have at least a few, albeit usually of limited size and potential. There may also be emphasis on awarding contracts to companies from developing countries to encourage local production or expertise. This may also arise due to an urgent demand when the company concerned is a representative of one in an EEC country. Contracts may be divided in lots for ease of allocation and to encourage local bids. Despite the emphasis on competition, cooperation between companies is also encouraged.

Enlargement demands on EDF

The UK is committed to the enlargement of the Community and endorses applications for membership by Spain and Portugal. Negotiations with Spain and Portugal could be lengthy. Their membership and the recent membership by Greece could further influence construction in Mediterranean countries and in developing countries, particularly through the use of the EDF which, although an important feature of EEC policy and certainly of great help to developing countries, should, nevertheless, be seen as a part of all multilateral aid, the growth of lending agencies and the greater influence of regional associations. All are likely to need the support of construction industries. In many ways the EDF emphasizes a changing role for all members of the construction industries of the UK and other EEC member states, the basis of which has been emerging gradually in recent years as a result of industry, financial and other pressures, and has led to private sector joint ventures, some dependence on effective central planning and private and public sector cooperation. There seems every reason why these partnership trends should now encompass supranational planning and cooperation, and, in view of the increasing cost of most construction, they may be essential.

Conclusion

In most world markets there is a constant reassessment of construction needs. The potential in developing countries, particularly those with mineral, energy and other resources which are essential for industrial countries is massive, but pursuing such markets is costly and increasingly competitive. Consultants who contribute much to UK earnings and prestige overseas face growing competition from those in Europe, the USA and elsewhere as they find it essential to replace lost or satisfied markets. Most find greater competition from designers in countries that previously relied on external expertise. Manufacturers of many UK products needed for construction continue overseas activities against competition which is unlikely to decrease. This may be the result of local incentives for the creation of new industries, often with lower costs and cheaper freight. There is a growing number of overseas designers, manufacturers and contractors who have learned their skills from UK firms and those from other industrial countries. They, too, are finding that local markets are changing, and are extending their sphere of activities, thus aggravating competition.

More and more those in the UK construction industry must rely on management and financial skills and the ability to undertake contracts involving advanced technology, often of considerable value which stretch financial resources, to be successful overseas. Most industrial countries see the encouragement of their construction industries overseas as a means of helping sales of manufactured products thereby benefiting home revenue and employment. Despite changes in markets it seems unlikely that local designers, suppliers and contractors will completely replace the advanced skills available from the UK and similar countries in the immediate future, however, their growing strength could mean that competition will increase. While technology advances so, too will the construction needs of each sector. Industrial development, that of the infrastructure and buildings like hospitals and hotels, rely on advanced equipment. Transport needs such as airports, railways and communications such as telephones and telex need expensive products and assembly. It is on these and maintenance skills that UK success could depend in the future, and which are a part of markets likely to provide essential contributions to UK trade and to that of others participating in overseas construction markets.

Index

171